Marlone D   ·.(

Glandular Failure-Caused Obesity

and Other Associated Promoters

William B. Ferril M.D.

The Bridge Medical Publishers

For Mckenna, Conner, Hayes, and Billy

The Bridge Medical Publishers

© 2004 William Ferril

Published by The Bridge Medical Publishers
P.O. Box 324
Whitefish, Montana 59937

Printed by Satori
200½ Wisconsin Ave.
Whitefish, MT 59937

Cover Photo by Synergy

ISBN 0-9725825-1-7
Printed in the United States of America

10 9 8 7 6 5 4 3 2

Acknowledgments

To my patients, who always point the way to better methods.

To the holistic mental giants that laid the clues for this book: Robert Atkins M.D;, Jeffery Bland, Ph.D.; Alan Gaby M.D.; William McKenzie Jefferies M.D.; John Lee M.D.; Uzzi Reiss M.D.; Andrew Weil M.D.; and Jonathan Wright M.D. To my wife, Brenda, who loves me in a way that facilitates my passion for holistic healing.

To my editor Hanna Plum for her patience and guidance.

To my layout and design expert Leana Schelvan.

To Ginny Wilcox, as usual, for helping to do the final details at Sartori Publishing.

Special thanks to Rick Nagle, Esq., for his professional instruction in regards to format for some of the earlier versions of this material.

Perpetual thankfulness goes to Steven Small Salmon who patiently instructed me about the futility of being seduced into the negative emotions.

To Patty Perigo for her kind heart and genius

Disclaimer

This book is intended as an educational tool to acquaint the reader with alternative methods of understanding, preventing, and treating obesity. The Bridge Medical Publishers hopes that this book will enable the reader to improve their wellbeing, and to better understand, assess, and choose the appropriate course of treatment. Because some of the methods described in this book are alternative in nature, by definition some of them have not been investigated and/or approved by any government or regulatory agency.

The information contained within *Glandular Failure-Caused Obesity* is not intended as a substitute for the advice and/or medical care of a physician. Nor is the content of this book intended to discourage or dissuade the reader form the advice of his or her physician. The contents of this book are for informational purposes only and are not intended to diagnose or treat illness. The diagnosis and treatment of illness requires a specific exam, tests, history, and an appreciation for each individual's uniqueness of presentation. For those, the reader is advised to seek the counsel of a competent holistic physician. Individuals who desire weight loss, improved health or healing from chronic disease need close supervision and follow-up from their physician.

If the reader has any questions concerning the information contained in this text, or its application to his/her particular medical profile, or if the reader has unusual medical or nutritional needs or constraints that conflict with the advice in this book, he or she should consult his or her physician before embarking on any medical treatments advised in this text. Pregnant or nursing readers should consult their physicians before embarking on the nutrition and lifestyle programs suggested in this text. The reader should not stop taking prescription medications without the advice and guidance of their personal physician.

Glandular Failure-Caused Obesity

Table of Contents

Introduction 10

Chapter 1
The Glandular Component 16

Chapter 2
Mineral Imbalance and Obesity's Creation 44

Chapter 3
Exercise and Obesity 59

Chapter 4
More Glandular Secretions That Affect Body
Weight 71

Chapter 5
The Torture Chamber Diet 81

Chapter 6
Syndrome X: 103

Chapter 7
The Last Three Hormones of the Basic Seven That
Centrally Influence Obesity 116

Chapter 8
A Deeper Understanding of Steroid Tone and
Pressure 128

Chapter 9
Obesity Has a Component of Causality in Poor Nutrition 152

Chapter 10
A Trip Down the Digestive Tube 197

Chapter 11
The Liver 214

Chapter 12
Obesity-Related Blood Vessel Disease Resulting from Nutritional Deficiency 245

Chapter 13
The Fear of Real Hormones and the Reassurance of Fake Hormones 253

Chapter 14
The First Day of Your New Life 258

Appendix A 264

Appendix B 273

Bibliography 285

Index 297

Introduction

I inherited the fat making genes. I decided that I would fight off my genetic inheritance. My success involves my recent education about what science truly knows about the making and losing of fat. The consequence of living within a profit driven health care system makes this scientific information largely inaccessible in its congruent whole.

Profit considerations influence which universities receive the pharmaceutical companies' research funds. This financial reality in turn influences the way medical schools educate physicians to think about disease. Of course physicians like myself have very little reason to doubt our educations until our patients begin providing living inconsistencies about how the body heals. Some physicians become tired and invested in the complex's way of doing things. I am living proof of how kind and informed patients can point the way to better methods.

The better methods kept growing and my way kept shrinking until I finally took a long sabbatical, and began to study what the medical textbooks really say about obesity and other middle-aged diseases (diabetes, heart disease, high blood pressure, and hormone imbalance). Almost four years later, I am truly shocked about the disparity regarding the sensationalism of what sells and what heals. I maintain that I am not any smarter than the average physician. However, I may have more curiosity than most about inconsistent details. Something inside me drives me to solve the inconsistencies that real live patients provide. This book compiles how one solves the obesity puzzle by putting all the pieces into a logical sequence. Once we become aware of the order to the 'pieces' of obesities creation focus will allow healing. Diets do not work. Awareness coupled with focus does work.

Obesity: Deactivating the Fat Making Machinery

Seven factors activate the fat making machinery:

1. The glandular component
2. Reversed mineral ratios diet component
3. Lack of exercise

4. Unmanaged stress
5. High carbohydrate diets (the torture chamber diet)
6. Vitamin and trace mineral deficiencies (nutritional deficiencies which propagate fat accumulation and impede its usage)
7. The psychological component (the pleasure of eating poorly)

Overview of the Problems Associated With Other Popular Diets

The Glandular Component

Six glands control the amount of body fat:
1. Pituitary
2. Thyroid
3. Liver
4. Pancreas
5. Adrenals (adrenaline and steroid compartments)
6. Gonads (the ovaries or testicles)

The big problem in America is that most overweight owners have not had these glands' function checked in a scientifically sufficient manner. Obesity always has a glandular component to its causality. Until the glandular component is identified and corrected the owner stays stuck within the torture chamber of worsening obesity.

Popular diets available today are incomplete for various reasons. The first problem concerns their incomplete attention to the fat making hormones. Successful weight loss requires consideration of seven basic fat maker related hormones - insulin, cortisol, androgens, estrogen, insulin-like growth factor type one (IGF-1), thyroid and epinephrine. In addition, indirect but important influences occur from the amount of growth hormone and prolactin secreted by the pituitary. One needs to consider these hormones first before starting any weight loss plan. Imbalances in the pituitary secretions cause disproportioned amounts in the basic seven hormones listed above. Obesity propagates as the result of inappropriately proportioned relationships between these basic seven hormones.

11

Weight loss becomes prevented and fat production becomes promoted when these seven hormones imbalance.

Theoretically, there are even more fat-maker related hormones. The jury is still out on how they really relate to the overall fat making machinery in the body. Rather than add these mysterious and so far without weight loss benefit hormones to the discussion, the tried and true hormones listed above will be the focus of this book. Meanwhile, the academicians can fight about these theoretical and peripheral hormones for gaining and losing fat.

Popular high-protein diets fail to manipulate all seven basic hormones that are a determiner of obesity. A successful diet must affect the hormone-based urge to consume food. In addition, it must inhibit the fat making hormones and encourage the fat burning hormones. The power of these hormone messages constitutes an important concept for weight loss and other health successes.

High protein/good fat and low carbohydrate diets improve the probability for weight loss by partially addressing hormone imbalance. Eating protein and good fats instead of carbohydrates reduces insulin – the fat maker. Using this principle also turns down the appetite center in the brain. The seven basic hormones for fats creation and the glands that secrete them are explained in chapters one, four and seven.

The Mineral Component

However, high protein diets often fail by creating a mineral imbalance between sodium and potassium. When mineral imbalances occur, weight loss curtails because the body then requires increased insulin secretion for even a small amount of carbohydrate intake (explained later). In this case, the failure rate becomes secondary to overlooking the importance of mineral balance. Mineral balance and its effect on weight loss are explained in chapter two.

The weakness in the high protein diet turns out to be the strength of some of the other diets. The high fruit, vegetable, and unprocessed grain diets often have superior mineral balance content. These diets fail because of the other hormonal imbalances they perpetrate in the already obese owner.

The Stress-Filled Lifestyle Component

Chronic mental stress provides another fat maker messenger. Unlike physical stress, mental stress causes an increase of insulin within the body. Until the obese owner gets counseling on the ways to sidestep this hormonal havoc, weight loss will be prevented. How stress affects fat accumulation will be explained throughout the manual.

The Movement Component

Physician's Sidebar

Ways of correcting these diet defects will be explored throughout this manual. The ideal diet combines the best of each diet and eliminates the part the sucks the owner back into the torture chamber. Understanding how the seven different body hormones either help or hinder weight loss is essential. Later, the mineral needs of the body will be explained. Implementing these inter-relating factors allows overweight owners to start on their healing paths.

Weight loss success cannot occur without an increase in body movement. Move or die. The body was designed to move around and resist the forces of gravity. Sedentary owners succumb to the shrinking forces that lead to old age: little muscles, little organs, weakened bones and loosened skin. The same lifestyle habits that lead to shrinkage in the above organ systems also contribute to an increase in body fat. Sedentary owners doom themselves to defeat until they assimilate an active lifestyle into the equation. The muscle versus fat chapter explains this important relationship.

The Nutritional Deficiency Component

Another component of the successful dietary approach concerns the necessity of certain vitamins for the removal of body fat. The science is all there but it paradoxically collects dust. One of the reasons it remains largely ignored arises from the disjointed and circuitous manner that it presents as within the

13

medical textbooks. Until these nutritional facts are organized in a way that both doctors and patients can understand, fat will continue to accumulate from nutritional deficiencies.

Ironically, America is the land of the nutritional deficiency diseases: obesity, heart disease, high blood pressure and diabetes.

The Psychological Component

The final component for obesities creation concerns the psychology of obesities perpetuation. Stepping outside the torture chamber causes stress and anxiety. The comfort of the chamber even though it tortures is better than the unknown of feeling healthy and fully living. The last chapter touches on this dilemma and provides further reading suggestions for how to conquer this ongoing problem.

Each of the above-listed component's contribution to making body fat is reviewed in the chapters that follow. Owners that understand the interconnectedness of these components are empowered to heal. Diets fail but insight heals. Healing often involves nothing more than a heightened awareness of how one became fat and how one loses fat. Awareness allows focus. Getting fat did not occur overnight. Likewise, shedding fat takes awhile. Most owners that follow the program outlined below will shed between fifty and one hundred pounds in the first year. In addition, they will gain back muscle, organ size, bone mass and skin health. In each success case it all started with a brave first step into a new life of healing. Many other steps of awareness follow the first step. This book guides by creating awareness for an individual's solution to heal their weight problem.

Ground Rules:

- I call anyone in possession of a body an owner
- I purposely use unique and graphic medical descriptors to facilitate lay person comprehension of medical knowledge
- The first chapter contains many single sentence facts that stand-alone. This initial format helps to build the reader's knowledge base.

- There are physician's sidebars throughout the text. These are designed to provide the extra information that physicians need before they can believe. Anyone can try to read these, but do not feel overwhelmed if they are beyond comprehension.

Chapter 1

The Glandular Component

Gland secretions play a powerful role in the creation of body fat and constitute the starting point for a successful weight loss plan.

Glands secrete information into the blood stream. The types and amounts of these various secretions determine how the body's 100 trillion cells spend their energy.

Optimal secretions facilitate the right amount of 'raw blood fuel' (amino acids, fat, and sugar), cell nutrition and cell function.

Poor secretions facilitate manufacturing increased fat, decreased organ function and accelerate the aging process.

The glands, listed above, secrete various hormones that affect the gaining and losing of fat. Hormones carry information to the body cells via the blood stream. The blood stream contains a 'sea' of information that changes, as the body needs change.

All hormones message content (information) concerns how the cells direct their energy expenditure.

Obese owners have the problem in one way or another of too much body energy being directed into the storage of energy. In addition, because obese owners have too much food energy directed into storage as fat, their cells are always hungry. These owner's cells will remain hungry until their hormones are changed into a message of fuel availability. Available body fat becomes fuel that combusts in the cell power plants.

The body stores energy mostly as fat.

The ideally weighted person stores about 80,000 calories as body fat. However, the obese person stores many times this amount.

No matter what the body weight is only 2,000 calories can be stored as sugar within the liver and muscle cells as glycogen. A cognizance of these relative amounts of storage abilities between these two fuel groups helps to elucidate the fundamental fat making problem. Many pinheads pontificate the platitude about losing weight being a simple matter of restricting calories. Oh, if it were only that simple. Hormones direct whether fuel is stored or fuel is burned up in the cell power plants. By definition, pinheads do not remember their basic science.

Losing weight requires attention towards redirecting fuel energy away from storage as fat.

The hormones giveth and the hormones taketh away. Hormones direct how the body treats body fuel (sugars, proteins, and fat). The right hormones within, minute to minute, create a balance between storage and combustion of these fuels. Healthy owners always have hormone balance. Unhealthy owners always have hormone imbalance. Ironically, American health care often overlooks this basic fact for how one becomes fat.

Until the quality of the hormones within improve weight loss efforts end in defeat.

Fortunately, out of the more than 100 body hormones only seven types centrally affect the making and losing of body fat. Two other hormones influence the central seven.

The glandular composite hormone report card is essential at the beginning of any weight loss effort.

The initial hormone report card includes:

- Fasting insulin and C-peptide levels
- Fasting insulin like growth factor type 1 (IGF-1) levels
- Adrenal steroids (including aldosterone)
- Androgen type gonad steroids (from the ovaries or testes)
- Thyroid hormones' levels (including reversed T3)
- Estrogen status for both men and women
- Epinephrine urinary output (metanephrine and normetanephrine) alternatively obtain a homocysteine level
- Prolactin level
- Growth hormone levels (surmised by looking at the amount of androgens and IGF-1 levels)
- Hemoglobin A1C

Each of the first seven hormone types has a specific role in the overall message that directs the manufacturing of fat (energy storage). Prolactin has special fat making properties through its effect on some of the first seven hormone types above. Lastly, growth hormone directly influences the amount of IGF-1 released. In practice it makes more clinical sense to measure IGF-1 levels instead of growth hormone levels.

Another fact about some of the obesity-causing hormones becomes important. Specifically steroids and thyroid hormones are among the most powerful class of body hormones. I call these select few most powerful hormones the level one-type hormones.

These most powerful body hormones, with the addition of vitamin A, are the only body hormones that possess the ability to directly instruct the DNA (genetic programs) within the 100 trillion body cells. All other body hormones cannot directly instruct the DNA programs within cells if at all. What the DNA programs (genes) do determine whether or not body cells spend energy wisely or unwisely. Obesity results from the unwise use of available body energy. The hormone message content that the cells receive determines body energy usage. Message content (information) to all body cells conveys by the body hormones via the blood stream. The blood stream can be thought of as a sea of information that changes, as the body's needs change. Bad information within the blood stream results from the wrong hormones secreting. Hormones direct how the body spends or saves energy. Too much energy storage message content results in excessive fat.

Many of the obesity-causing hormones are amongst the most powerful type because they uniquely direct the DNA programs into activity or silence. The activity or silence of the DNA programs within one's 100 trillion cells powerfully determines calorie expenditure. DNA program activity also determines repair versus disrepair within a cell. Part of the reason owners gain fat results from their most powerful hormones, which instruct their DNA, being imbalanced.

Here lies the central explanation for why obesity associates with an accelerated aging rate: The hormone mismatch, which either allows or creates obesity are among the most powerful body hormones.

Physicians' Sidebar

Vitamin A and its Steroid-like Properties

By its name, one immediately misunderstands vitamin A (retinol). Unlike most other vitamins, vitamin A contains message content by virtue of its molecular shape. In addition, vitamin A has the ability to go anywhere within the body and deliver message content to the DNA program of a cell. In contrast, other vitamins work by facilitating chemical reactions within the cells. Vitamin A behaves more consistently as a hormone. In contrast, by calling it a vitamin, this creates a tendency to hit an intellectual roadblock that needs to be crossed in order to appreciate the consequences of this molecule's deficiency or excess. A "real food" diet (see the torture chamber diet chapter) would supply ample vitamin A and little tendency to develop deficiency of this important substance.

Vitamin A must be obtained in the diet and can become toxic, at high levels, because at high levels it amplifies its instructional content of cellular DNA beyond healthful

19

parameters. The only way to overdose on vitamin A results from ingesting high dosage supplements (above 50,000 IU a day or about two carrots a day) for over three months. Some individuals can ingest much higher amounts without toxicity. After all 50,000 IU's equals only 50mg. Too much vitamin A causes thinning hair, dry and scaly skin, bone spur formation, and brittle bones.

Too little vitamin A leads to diminished functional abilities of cells that coat the body (skin and cornea) and cells that line body cavities (the gastrointestinal tract and lungs). Conditions like ichthyosis vulgaris result from vitamin A deficiency. Skin cancers are promoted by this deficiency as well. Yet, little encouragement comes from physicians for decreasing cancer risks by taking adequate vitamin A. Adequate adrenal function depends on vitamin A to instruct adrenal cell DNA activation programs. In the healthful state, the liver fills up with vitamin A and releases it as needed.

Vitamin A denotes a complex of similar vitamins that promote cell maturity. Immature cells form a central property of cancer cells. Adequate vitamin A complex intake provides a cornerstone for the prevention of cancer. Almost all cells contain DNA programs that respond to the message content of vitamin A complex. Some vitamin A derivatives occur in plant seeds. These prevent cell division until their removal when the right conditions occur. For example the right conditions occur when the planting technique is correct. Likewise, within the body, Vitamin A's message content involves keeping the cells from undergoing rampant cell division. This fact continues to be almost entirely ignored by mainstream cancer specialist in their attempts to stop cancer cell division (tumor growth).

Lastly, the thyroid hormone properly activates the DNA programs within the body's 100 trillion cells only if sufficient vitamin A message content occurs. Some clinically low thyroid function patients therefore arise solely from vitamin A deficiency. When this causes the clinical symptoms of low thyroid function the thyroid test routinely done at the doctor's office will come back normal!

When the most powerful body hormones are out of kilter, obesity and other health problems are encouraged because body energy becomes misdirected.

For example, instead of rejuvenation hormone message content directed at cell repair processes being heard by body cells excessive body energy channels into to fuel storage activities (fat manufacture).

This fact partly explains why obesity increases the risk of heart disease, diabetes, high blood pressure, and arthritis (middle age related disease). Each of these diseases has a large component of causality in the abnormal hormones associated with obesity. This is good news because it means that by correcting the obesity-causing hormones it will also impact the middle age related diseases.

The basic seven hormone types that when imbalanced cause obesity (note: these are all glandular secretions)

- Thyroid hormones
- Insulin
- Insulin like growth factor type one (IGF-1)
- Cortisol
- Androgens
- Estrogen
- Epinephrine

The additional two pituitary secreted hormones, prolactin and growth hormone, which influence the amounts of the basic seven.

Thyroid Hormones

Thyroid hormones (T3 and T4) direct the DNA to increase the manufacture rate of cellular mineral pumps and furnace combustion chamber components.

The lack of thyroid message content within an owner's body diminishes the furnace flames within his/her one hundred trillion cells. The most reliable manifestation of low thyroid gland function is a low body temperature upon awakening.

A poorly functioning thyroid gland leads to a decrease in its message content to the numerous body cells. One component of the thyroid hormone message concerns its directions to invest in furnace component upgrades. Scientists call the furnace or power plant of the cell the mitochondrion. Poorly functioning mitochondrion are analogous to heating one's home with a furnace with worn out components and plugged air filters. For the same reasons that these homes remain cold, an owner with worn out mitochondrion components in his cells stays cold.

Most body cells prefer fat as their fuel source.

All body cells need certain key vitamins to process raw body fuels (protein, carbohydrate, and fat) into a processed fuel called acetate. Acetate is the simplest fatty acid. No matter what the raw fuel (protein, sugar or fat) consists of before it can burn up in the cell power plant it needs refining into acetate. The different raw fuels need specific and numerous vitamins to process into the refined fuel, acetate. Many owners have weight gain simply because they lack the necessary vitamins to turn raw fuel into refined fuel (explained in nutritional deficiency chapter).

The cell furnaces can only burn acetate within their combustion chambers. This last fact is analogous to the requirement for refined fuels within power generators of the physical world. For example, automobile engine design usually requires the refinement of raw oil into gasoline. In general, the combustion chamber of each power generator is designed to work only with a certain fuel type.

Similarly, the numerous body cell furnaces can only combust one fuel type. Therefore many thyroid problems are made worse by specific vitamin deficiencies that impede the processing of raw body fuels into acetate. A diminished delivery of the refined fuel, acetate, will exaggerate weight gain because it tends to also contribute to hungry cells (see nutritional deficiency caused obesity chapter).

Thyroid hormones deliver an additional message to the body cells beyond its heat creating effect within the numerous cell furnaces. The additional message involves its ability to direct the creation of membrane pumps that trap some of this heat energy for useful work. Again this is similar to a car engine creating heat but some of the energy is used to move the car down the road by engaging the drive train. Power plants also lose energy to heat but again some of this energy is trapped in the form of electricity. The body cells also operate on an electrical system. A cell's ability to charge its membrane proves critical to life. The energy needed to charge the cell membrane derives from the energy trapped within the mitochondrion combustion chamber in the form of ATP. The more ATP available determines one aspect of how high a cell can charge up its membrane. Membrane pumps charge the cell membranes. Membrane pumps need the energy contained in ATP to pump up the electrical charge (increase the voltage) contained in the cell membrane. The constant supply of ATP needs to occur or the electrical charge contained in the membrane diminishes and the cell dies.

Thyroid message content provides the informational direction that tells specific DNA programs to activate. In turn, these activated and specific DNA programs direct the manufacture of more membrane contained mineral pumps. The GA zillions of membrane pumps within the membranes of all cells of a body are made from proteins, which are coded for by the specific DNA programs mentioned above.

The more mineral pumps within the cell membrane then a additional determinant of higher membrane electrical charge becomes satisfied.

The greater the 100 trillion body cells' cumulative electrical charge is the more calories burned to maintain it.

In fact, the cell membrane mineral pumps comprise the reason that most calories burn up within the body. Fewer mineral pumps lead to fewer calories burned. Fewer calories burned leads to a lower metabolic rate.

Body cells utilize the electrical charge within the membrane to perform the cellular work of living. The more work performed then the more calories combusted. This is analogous to the charge within the membranes of a car battery, which can be drawn upon to power the electrical gadgetry within the automobile. Similarly, the cell power plants recharge the cell membrane like the cars engine recharges the car battery.

All body cells protect themselves and perform work by utilizing the energy contained in their membranes. Fully charged cells are maximally alive. Fully charged cells burn more energy (fat). The more energy consumed, the more calories consumed. Later, in the mineral chpater, it will be explained how the body cells cannot fully charge without the proper proportions of mineral intake within the diet (the third determinant of cell membrane charge).

For now, it is important to summarize the two main messages that the thyroid hormone conveys to many different body cells. 1) Carries to the DNA the message to invest in furnace component upgrades. 2) Carries to the DNA the additional message to create more mineral pumps within the membrane, which charges the cell electrically. Only when both of these processes occur smoothly can a normal amount of calories burn.

Certain DNA programs need the additional presence of sufficient vitamin A message content to activate what the thyroid hormone message started. In this way, vitamin A and thyroid hormones work synergistically.

Insulin Carries the Fat Maker Message and IGF-1 Competes with the Fat Maker Message

Insulin is made and secreted by the pancreas

- IGF-1 within the blood stream is made and secreted by the liver

There are only two hormones that allow the cell 'fuel tanks' to fill up. It helps to think of these two hormones as fuel nozzle hormones. Just as in filling a car with fuel, a fuel nozzle is necessary, so it is with most body cells, a molecular 'nozzle' is necessary to fill the 'tanks' of the cells with the fuel circulating in the blood stream.

The two 'fuel nozzle' hormones are

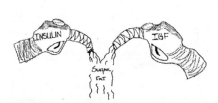

insulin and insulin like growth factor type 1 (IGF-1). Healthy people have at least 100 times more IGF-1 compared to insulin in their blood streams. Each of these two fuel nozzle hormones has a preference for the cell types it prefers to fill up. Because there are roughly 100 times more cells that prefer IGF-1 it makes sense that healthy people have at least 100 times this hormone as compared to insulin. Only when the body becomes unhealthy and IGF-1 levels consequently fall off will the body attempt to raise insulin hormone production rates so as to keep the total amount of fuel nozzle hormones constant. Hungry cells result when fuel nozzle hormones become scarce.

However, the insulin hormone contains message content that the IGF-1 hormone does not share. Insulin contains the fat maker message. Overweight people have a problem with too much fat maker message. Until the fat maker message decreases they will continue to gain weight year after year.

Body fat is not possible without enough insulin message content to maintain it.

The other six hormone types of obesity, when unbalanced, facilitate the amplification of the insulin message.

Healthy people have almost no insulin in the fasting state but have high amounts of IGF-1. Unhealthy and obese people have higher fasting insulin and/or C-peptide and lower IGF-1.

Therefore at the start of any dietary program obtain these values for a baseline.

Insulin's major message concerns fuel storage.

Consistent with this fact is the increase in insulin following meals. The majority of the insulin message directs the liver to store up to 400 grams of glycogen within the liver. When the liver already has enough glycogen stored the insulin message then directs the liver to convert the additional sugar into fat and cholesterol.

The opposite situation occurs between meals because blood fuel tends to drop off. In these situations fuel needs to be released from the liver and not stored. New supplies of fuel need to be present in the blood stream to constantly provide nourishment to the body cells. Blood fuel content delivery, into the body cells, is not possible without adequate fuel nozzle hormones available to facilitate this process. In healthy people, when the body's cell fuel levels begin to fall off, the resupply of fuel occurs via the other fuel nozzle hormone called insulin like growth factor type 1 (IGF-1). IGF-1 occurs at levels greater than 100 times that of insulin in healthy people. This hormone's design helps the body cells outside the liver and fat cells to procure the fuel they need. IGF-1 levels, within the blood stream of healthy owners are 100 times insulin levels even following meals because its half-life is four days. In contrast insulin's half-life is about ten minutes. IGF-1 levels increase slightly between meals, while exercising and in the fasting state. Two other hormone types and two lifestyle habits need to occur for adequate IGF-1 levels to be maintained beyond middle age.

The two hormones are: growth hormone and androgen. Growth hormone releases from the pituitary gland, when blood fuel levels fall off. Growth hormone release tells the liver to release more IGF-1 and fuel (sugar and fat). Androgens, like testosterone and DHEA, tell the liver DNA programs to activate towards the manufacture of more IGF-1. So androgen causes IGF-1 creation and growth hormone causes IGF-1 release. Normal blood stream IGF-1 levels require both normal growth hormone release and androgen levels (testosterone or DHEA).

Sufficient androgens, like DHEA and testosterone, direct the liver to make adequate IGF-1. There are others but for now these two provide the basics.

Sufficient growth hormone secretion causes the liver to release the manufactured IGF-1 into the blood stream

The lifestyle habit concerns the fact that the blood fuel falls between meals (fasting) and when one exercises. Consequently, these two habits encourage the release of growth hormone. This knowledge explains why chronically underfeeding (fasting) both in animals and humans results in a lengthened life span. The underfed human will tend to secrete more growth hormone. Increased growth hormone arriving at the healthy liver chronically will help promote elevated IGF-1 levels. Elevated IGF-1 levels provide the nutritional advantage to the organs, bones and muscles of the body. Whereas, elevated insulin provides the nutritional advantage to the liver where more fat and cholesterol manufacture occurs. Adequate IGF-1 in the blood stream facilitates the body cells outside of the liver in their procuring nutrition. Adequate insulin, but not too much, facilitates the liver and fat cells in storing fuel following a meal. Healthy owners, between meals and when exercising, draw down the stored fuel within their liver and fat cells, preventing obesity.

Unhealthy owners have a decreased growth hormone secretion rate and hence a diminished IGF-1 level. Death is prevented between meals when their blood fuel falls because two other hormonal aberrations save them. First, the stress response hormones release and activate catabolic pathways. Catabolic means that body structure dismantles to free up more fuel. The body structure is composed of sugar, fat and proteins. Second concerns the exaggerated amounts of insulin needed to help the body cells procure nutrition because IGF-1 levels have fallen. Remember, there needs to be a constant amount of total "fuel nozzles" around to fill the cell fuel tanks. Insulin in large amounts (insulin resistance) gets past the liver and into the general circulation where it acts much like IGF-1.

Unfortunately, when insulin secretes in large amounts the highest concentration still arrives at the liver where the fat and cholesterol making machinery preferentially become activated. High insulin and/or C-peptide in the fasting state provide a good laboratory marker for these types of obesity prone owners. Modern literature calls these people Syndrome X, Metabolic Syndrome or Insulin Resistance Syndrome (expanded discussion in chapter six).

Interestingly, the fact that these individuals have at the root of their problem a falling IGF-1 level is almost universally ignored. In its place peripheral approaches like cholesterol lowering drugs, drugs that whip the pancreas into even more insulin production and high blood pressure medication are prescribed. All these accepted approaches do little to heal the underlying problem and have side effects to these owners' health. Until steps are taken to raise these owners IGF-1 levels both obesity and accelerated aging will continue to propagate.

Healthy owners have the ideal balance between insulin and IGF-1. In these situations insulin increases following meals and the IGF-1 stays unchanged. Balance between these two hormones allows the liver to uptake a proper proportion of fuel compared to the amount that other body cells are allowed to procure with the assistance of IGF-1 following a meal. In addition, the fuel stored by the liver in a healthy person becomes readily available between meals, when exercise or a fast occurs. In all three of these situations, as blood fuel begins to fall, growth hormone releases, which directs the liver to release the stored fuel (sugar and fat) and more IGF-1 but not protein. The extra IGF-1 released further facilitates the hungry body cells uptake of nutrition.

Remember that only growth hormone prevents the catabolism of body protein for fuel between meals. Here lies the chronic weakening mechanism when unhealthy people are between meals, fasting or exercising. As their blood fuel falls in these situations the wrong hormones make body protein available for their fuel needs because growth hormone release proves inadequate.

Unhealthy owners also have diminished IGF-1, as a consequence of diminished growth hormone release, following meals and between meals. In both situations this sets up the need for abnormal secretion rates of insulin (commonly referred to as insulin resistance). Remember insulin contains the fat maker message. The higher the insulin secretion rate, the higher the fat maker message within. Syndrome X owners are doomed until their need for insulin lowers. One of the ways insulin needs fall off is when IGF-1 levels return to normal. IGF-1 contains none of the fat and cholesterol stimulating message content that insulin contains. Rather, IGF-1 message content concerns itself with the procurement of nutrition (fuel) by other body cell types, outside of the liver and fat cells.

It was briefly mentioned above that unhealthy owners have other hormone increases, which facilitate resupplies of blood fuel between meals. However, there is a price paid from this aberration to the body structure in the form of lost muscle and organ mass. This largely ignored medical fact explains another consequence of the unhealthy body's lack of sufficient stimulus for growth hormone release. Only when adequate growth hormone release occurs will there be a sufficient counter message to prevent protein destruction for fuel needs. Muscles are made from protein. Growth hormone's presence prevents the combusting of protein when blood fuel levels fall. Instead, growth hormone preferentially directs the liver and fat cells to dump sugar and fat into the blood stream while sparring protein. This fact partially explains why healthy owners have nice muscles while unhealthy owners do not.

Physician's Sidebar

The Myth of Insulin Logic

Mainstream medicine's propaganda dogma grooms physicians and patients to focus on insulin as the nutrition uptake hormone. Meanwhile they ignore the scientific fact that IGF-1 occurs at levels greater than 100 times those of insulin in the healthy individual. The disconnection between this simple optimal ratio maintains itself by arbitrarily measuring insulin in micro units and IGF-1 in nanograms throughout the medical texts. More disconnection results from the fact that numerous different names describe IGF-1. For example, the older medical literature describes IGF-1 in the following additional three ways: non-suppressible insulin like activity of the blood, sulfation factor, and somatomedin C. In order for a physician to appreciate the important role that IGF-1 plays in the cells in obtaining nutrition he/she would need to be aware of and have time to look up all four of these alternative descriptors for the same hormone. Reconnecting all four different names with the facts associated with them and a common measurement method allows the important role of IGF-1 to emerge.

Simple logic shows that there is not enough insulin to go around to all body cells. One liver or fat cell contains 200,000 insulin receptors. The 100 times more IGF-1 compared to insulin makes up the volume discrepancy needed to deliver fuel uptake message content to the other body cells. Remember, whatever the insulin secretion rate, the liver because of its portal vein connection to the pancreas, will always receive the highest concentration of insulin message content.

The basic sequence of protein destruction, when unhealthy owners find themselves between meals and their blood fuel begins to fall, results from other hormones.

The entire group consists of glucagon, cortisol and epinephrine. Like growth hormone they all facilitate the release of sugar and fat into the blood stream. Unlike growth hormone body protein dismantling (gluconeogenesis) becomes fair game when growth hormone levels become insufficient. Here again, another vicious process destroys the physiques of countless middle-aged owners. The importance of protecting their body protein content needs to be explained to them.

One of the central determiners of body protein content relies on adequate growth hormone levels for the above-stated reason. The other consequence of a falling growth hormone secretion concerns the fact that less IGF-1 releases into the blood stream from the liver. Less IGF-1 release means that the optimal 100:1 ratio between IGF-1 and insulin diminishes. When IGF-1 falls then insulin levels must rise to facilitate a way for the other body cells to procure nutrition. Remember that the fuel nozzle total hormone levels, between insulin and IGF-1 needs to stay constant or the body cells become hungry. Therefore a fall in IGF-1 necessitates a corresponding rise in insulin to keep the total fuel nozzle hormones level the same.

Anatomy of this imbalance: portal vein connection between the pancreas and the liver. Secondary to this relationship the liver will always receive the highest insulin message content for a given secretion rate from the pancreas secreting into the protal vein, which leads directly into the liver.

Remember, the insulin message concerns itself with storing food energy in times of plenty (following a meal). Therefore, the higher the insulin, the more that body fuel tilts towards storage sites (liver and fat cells).

In healthy owners, some insulin is necessary to allow enough fuel storage following meals in order to provide fuel in the between meal state (fasting and exercising). However, the 100 fold greater amounts of IGF-1 present in healthy owners following meals tips the blood fuel towards the body cells outside the liver and fat. This optimal ratio facilitates the sufficient nutrition of the cells throughout the body. Hungry cells occur when insufficient amounts of either insulin or IGF-1 occur. In addition, when insulin must increase to compensate for a falling IGF-1 then the fat making machinery abnormally activates.

The additional troubles within obese owners, such as other hormone abnormalities, nutritional deficiencies, dietary indiscretions and/or mineral deficiencies all diminish IGF-1 levels as well (discussed later). The fall in IGF-1 levels means that the peripheral cells have a decreased ability to uptake nutrition following meals because of the anatomical secretion pathway of insulin (see figure). This anatomical fact results in the liver always receiving the highest concentration of insulin's message content. Abnormally high insulin levels with low IGF-1 levels means cells outside the liver have a decreased ability to obtain nutrition and fuel. Instead the liver, in these cases, obtains the lion's share of body energy and directs it into the storage pathways (fat).

This means that because insulin levels have increased their livers make more fat and cholesterol for the same amount of caloric intake.

Healing obesity involves facilitating a rise in IGF-1 levels. Weight can then be lost as insulin needs fall off.

Insulin has a half-life of ten minutes. IGF-1 has a half-life of four days. Because healthy people have adequate IGF-1 levels while fasting or between meals, they have no need for insulin during these times.

Here lies another valuable clue: why would a body elevate insulin levels in the fasting state?

By definition an elevation of fasting insulin levels is either called Syndrome X or Metabolic Syndrome. The fact that IGF-1 levels diminish in this syndrome is almost completely ignored. Instead, usually a peripheral approach results in prescribing blood pressure medication, cholesterol lowering medication, and insulin increasing medication, which treats only the symptoms of this disease process.

Insulin increases brought on by some diabetes drugs only worsen the obesity problem. The pancreas design was not intended to produce massive amounts of insulin to shore up falling IGF-1 levels. The pancreas is little and the liver is large. IGF-1 at amounts one hundred times insulin is easily produced and secreted by the large liver. The pancreas however has to work itself to death when IGF-1 levels fall even a little bit (example: IGF-1 falling only five percent means the pancreas has to work five times harder). The little pancreas was not designed to increase insulin production many fold in order to shore up a falling IGF-1 level. These owners eventually can exhaust their pancreas (beta cell burn out) and this leads to one form of diabetes (see liver chapter).

Simplistically, two hormones determine the amount of IGF-1 available and released (there are others but for now the all important first two are discussed). First, the amounts of androgen steroids like testosterone and DHEA secreted. Second, concerns growth hormone secretion rates.

Basically, growth hormone gets secreted when the brain senses that the blood fuel level has fallen. Blood fuel levels tend to fall with two situations:

- Fasting (between meals)
- Exercise

The trouble with a sedentary lifestyle and obesity concerns two things:

- Other associated hormone abnormalities prevent the fall in blood fuel levels
- Exercise does not happen

Because of these two processes growth hormone secretion rates drop off. Elevated blood sugars caused by these other hormonal aberrations powerfully inhibit growth hormone release. Stress filled and sedentary lifestyles elevate the blood sugar, as well, and this inhibits growth hormone release for the same reasons (explained later). This leads to a decreased secretion rate of IGF-1. The fall in IGF-1 increases the need for more insulin. More insulin message content increases the production rate of changing sugar into fat and cholesterol. These are packaged into LDL cholesterol and move towards the storage depots in the abdomen and macrophages that line the arteries.

Scientific fact: high insulin levels turn on the cholesterol and fat making machinery.

Normal insulin secretion rates leads to a normal cholesterol and fat manufacture rate. This simple relationship is largely forgotten within the hysteria to sell more cholesterol lowering prescription medication. Cholesterol lowering medication has side effects. Some authorities liken these side effects to the acceleration of the aging process.

Androgen Steroids

Androgen steroids: made in the adrenals and gonads (ovaries and testicles). Their message content always concerns rejuvenation and repair of body cells. Two major androgen types, one secretes from the gonads and the other from the adrenals.

3. DHEA secretes mostly from the adrenals
4. Testosterone secretes mostly from the gonads

Remember that steroid hormones are among the most powerful hormone class. Only the most powerful body hormones directly instruct the 100 trillion cell DNA programs. Depending on the steroid type, certain DNA programs (genes) activate or

shut down. What the message involves for a given steroid type results from its precise shape. Gene silencing results in a cessation of protein synthesis for which that gene codes. Gene activation results in protein synthesis specific to the gene activated.

Proteins constitute the metabolically active component of body tissue. In other words, proteins when active burn up calories. Examples of metabolically active protein types are: enzymes, mineral pumps and cell receptors. This chapter begins the discussion of the first two types of steroid message content: anabolic versus catabolic.

The androgen message (anabolic) concerns itself with rejuvenation (cell repair and build up) activities.

The catabolic steroids deliver the opposite message. Catabolic message concerns itself with consuming body structure for the creation of energy. Body structure is made up of fat, carbohydrate and protein. High levels of catabolism make all three raw fuel sources fair game.

One facet of high levels of catabolic steroid's message concerns energy channeling for a perceived survival threat whether real or imagined. The stress response activates this system.

Body structure dismantles itself when the stress response activates in order to maximize available energy.

Stress steroids (the catabolic type) are only made in the adrenal glands. Both the adrenal and gonads manufacture the androgen steroids.

Healthy owners have a balance between survival message content and rejuvenation message content. Unhealthy owners are out of balance between these two opposing messages.

This subsection concerns the androgen message and how it curtails obesity. In the next subsection, the survival message (catabolism) elicited by stress will be reviewed in regards to its perpetrating obesity.

Only real hormone shapes carry accurate information.

Hormones carry information via their precise shape. Steroid hormones are relatively small and simple in the world of other much larger hormones. The smaller the hormone then the more the message will change with even a slight change in shape.

In order to obtain a patent advantage, drug companies must change the shape of the natural body hormone. Natural body hormones deliver precise messages. Unnatural hormones, such as being patented by drug companies, create profit but because their shape has been altered their message content changes as well. One prominent example concerns horse estrogen collected from pregnant horse mare's urine. There is some human type estrogen within this mixture of horse estrogens given to millions of women. The pharmaceutical companies cannot get a patent advantage without selling the whole mixture.

However, the mixture contains unnatural shaped estrogens that one does not find in the human body. The altered shaped estrogens deliver abnormal message content. The message content differs because the shape differs.The unnaturally shaped horse estrogens deliver an altered message to the ingesting owners trillions of cellular DNA programs. Uzzi Reis M.D.

Ob/Gyn and author of the book, *Natural Hormone Balance for Women* says it very well. "Are you a horse? Do you eat hay? Then why take horse estrogen?"

Estrogen's role in creating body fat will be discussed more completely in a following subsection. For now realize that estrogen promotes anabolism for fat cells in the body by a circuitous route. The higher estrogen becomes relative to androgen, the larger a women's breast and hips.

For now, it is only important to begin to appreciate that the different types of steroids need to occur in the proper amounts and timing or the message content reaching the DNA becomes altered.

Unnatural hormones carry inaccurate message content to the trillions of body cells. Unnatural hormones carry faulty information because their shapes deviate from the natural hormone's shapes.

When a steroid hormone's shape is changed to get a patent, then the message content changes.

Cells that receive altered directions on how to spend their energy manifest as side effects and disease eventually.

One of the side effects when the androgen message diminishes concerns the fact that the afflicted owner's body gets older. Bodies get older when their cells receive diminished instructions to repair and rejuvenate. Androgen type steroids carry the rejuvenation message content to the body cells. Remember, the androgen message concerns rejuvenation instructions within the trillions of cell DNA programs. It is the lack of rejuvenation message content that accelerates the aging process. Some components of the aging process are the loss of muscle and bone but also the gaining of body fat.

The androgen message contained in testosterone and DHEA shed fat in one important way. Androgens direct the liver DNA programs (genes) to synthesize IGF-1. High IGF-1 blood levels mean that one requires less insulin for their cells to procure their nutritional needs.

Less insulin means that the fat maker message diminishes.

Many women notice weight gain after their ovaries are removed. Some of this tendency can be explained by the loss of androgen message content. Healthy ovaries and adrenals make androgens.

Weak adrenals incompletely compensate for the removed ovaries. Menopause also brings on diminished ovarian function. Health after menopause depends on healthy adrenals, which produce significant androgens.

Many postmenopausal diseases have a component of causality in poorly functioning adrenals. Examples are: osteoporosis, obesity, and chronic fatigue.

A well run 24-hour urine test can check the thyroid hormones, adrenal steroids, and gonad steroids. Over a 24-hour period in a typical day the peaks and valleys of hormone release will average out. In contrast, the standard mainstream approach entails a blood drawn sample for measuring one or two of these important hormones. The instant of the blood draw only measures that instant in time of where the body directs its energy. Whether the values obtained prove the high or low for the day is not discernible with this approach.

For example, if the act of drawing blood causes stress, then the rise in stress hormones, like cortisol, causes the displacement of the bound thyroid hormone fraction and thus artificially increases the true average of this hormone in one's blood stream. Forgetting this fact doctors often falsely reassure low thyroid functioning patients that their thyroid tested normal when it is not. However, taking the time to measure thyroid hormone production over a twenty-four hour period averages out these peaks and valleys.

The volatility of the blood stream levels makes sense when one realizes that the hormones direct how the body spends its energy. The many different activities and situations of life demand different message content to direct body energy appropriately. The resting state predominantly provides a time when body energy directs into rejuvenation activities. The right hormones need to occur to oversee this process. Likewise, exercise demands different body hormones to direct the freeing up of energy that allows this activity to continue. A blood sample taken in either of these situations would show entirely different results.

Remember that thyroid hormone, all steroid hormones, and vitamin A are among the most powerful body hormones because only these directly instruct the trillions of cells' DNA programs. Only these most powerful hormones directly determine a cell's genetic program activity. The genetic program activity determines which proteins are made. The amounts and types of proteins made by a cell determine its productive capacity, overall integrity, and repair rate. All the many other body hormones cannot directly instruct the DNA programs, if at all.

Keeping these facts in mind it becomes more logical that the overall mixture of hormones within constitutes an important determinant of health. The overall mixture of hormones within determines the informational direction of how the body spends its energy. Some bodies spend energy wisely and there occurs a continuum all the way down to those bodies that spend energy foolishly.

Healthy bodies have proper proportions of these important hormones directing their cells in the wise use of available energy. The opposite situation exists, in that unhealthy bodies spend their energy unwisely because their hormones are wrong. The wrong hormones result from nutritional habits, lifestyle habits, patented forms of unnaturally shaped hormones and glandular failure.

All hormone message content concerns how the body directs its energy expenditure.

Healing an unhealthy body cannot occur until better informational direction occurs, via the hormones, on how their body spends its energy.

Remember, obesity has a component of too much message content directing body energy into storage. Hungry body cells result when too much energy channels into storage (fat).

Steroid tone provides a construct that helps quantify the quality of the mixture of blood stream information carried by these important types of body hormones. The steroids are among the most powerful body hormones because they each carry specific message content, inherent in their unique shapes, directly to the DNA. Each specific type's message individually direct the trillions of cell DNA programs (genes). Too much or too little of a particular steroid leads to either excess or deficiency for that message, respectively.

Healthy people always have the proper amounts of these most important hormones. This occurrence results in the proper amount of repair to rest ratios within the body cells. Middle age begins to occur when the quality of the information delivered to the cells by these most powerful hormones begins to deteriorate. One of the consequences of middle age with a large component of causality in the deteriorating message content concerns obesity.

Almost all the body cells have a complete genetic program. The genetic program activity determines cell repair rates, rejuvenation rates, functional ability and the overall integrity of the cell.

Paradoxically, mainstream medicine often overlooks the central importance, for continued youthfulness, of bodies possessing correct message content directing their cells.

Steroid tone is measured by a well-run 24-hour urine test. It provides a hormone report card for these most important body hormones.

The more optimal the amounts of the different steroids proportional to one another, the higher the steroid tone becomes. Adequate rejuvenation and repair activity occurring requires a sufficient amount of steroid tone. A high steroid tone signifies that there are proper proportions of message content among the various steroids. In other words, the catabolic message content balances with the anabolic message content.

Unhealthy owners will have poor hormone message content. Their health will continue to deteriorate until improvement to the quality of the message that directs their cells occurs.

Steroid pressure provides another useful construct that helps to visualize the behavior of steroids once released from their glands of origin.

Only the most powerful hormones (steroids, vitamin A, and thyroid) can access any body tissue or chamber. There are essentially no barriers to their penetration. The amount of their presence within the body periphery (the joints, bones and skin) is determined mostly by their initial amount of release from their gland of origin.

Like smoke in a room, which moves from a high concentration to a lower concentration periphery, steroids diffuse towards the periphery of the body kingdom.

A poor steroid generation rate at the source (the adrenals and gonads) diminishes the pressure head. It is the periphery tissues that suffer first. A falling steroid pressure manifests as wrinkled skin, painful joints and osteoporosis.

The periphery of the body kingdom is the most vulnerable because the presence of rejuvenation message content falls off here most severely because these tissues are furthest from the source of this information.

Steroid tone quantifies the quality of the steroid message content reaching the body cells. Different types of steroids have unique and precise shapes. It is the unique shape that conveys the message. Change the shape and the message content changes.

Steroid pressure quantifies the amount of a specific steroid reaching the body cells. For example, an adequate androgen message directs sufficient rejuvenation activities. In contrast, inadequate androgen message content leads the body's cells to fall into disrepair.

The signs of disrepair show up first in the periphery:

- The skin
- The joints
- The bones

Stress Hormones and Weight Gain

The stress hormones cortisol and epinephrine are both made and secreted by the adrenal glands. This section concerns the cortisol component of the stress response. High levels of cortisol, as seen in the stress response, consume body structure to maximize energy, a catabolic effect.

The first facet of the stress response and weight gain

Sadly, the association between the chronic activation of the stress response and its tendency to promote obesity is often not mentioned within the mainstream medical approach. Even though the fact is that when owners are sedentary and they continually experience mental stress, whether real or imagined, more insulin needs to be released. The more insulin released the greater the message content to create fat. How can this be?

The body cannot discern the difference between mental and physical stress. Therefore the physiological changes in both instances are the same. The physiology of the stress response largely involves outdated energy channeling when mental stress occurs. Even though mental stress does not need increased blood fuel (sugar, fat and amino acids) to survive it, the body responds by dumping fuel into the blood stream as if a physical challenge was forthcoming. In prehistoric times the extra fuel proved advantageous because the fuel allowed increased strength within the muscles for running from the jaws of some large animal.

Modern stress is largely mental in nature. The extra fuel released consequently has nowhere to go. The body eventually figures out that it has been fooled into dumping massive amounts of sugar into the blood stream and insulin secretes into the blood stream to rectify the situation. The increased insulin directs the activation of the fat maker message, which causes problems in the setting of chronic mental stress.

This fact explains why the old adage about "walking it off" has merit. Exercise causes the extra fuel within the blood stream to be sucked into the muscles and combusted for the creation of energy. Exercise also increases androgen production, growth hormone output and IGF-1 secretion rates. All these hormones combat the need for insulin (see above discussion).

The stress response and the second facet of weight gain

The major hormone of the stress response is cortisol. Remember that cortisol is from the steroid class called catabolic. Catabolic hormone's message content concerns the consuming of body structure for fuel release into the blood stream. Unhealthy owners have other hormone increases, as well, which facilitate fuel delivery with a price paid by consuming body structure between meals and when stressed. The price is paid to the body structure in the form of lost muscle and organ mass. This largely ignored medical fact is explained again by the fact that unhealthy bodies lack a sufficient stimulus or ability to release growth hormone. Only when adequate growth hormone secretes, from the pituitary, will there be a sufficient counter message to prevent protein destruction for fuel needs. Muscles and organs functional components derive from protein. Growth hormone's presence impedes the combustion of protein when blood fuel levels fall. Instead, growth hormone preferentially directs the liver and fat cells to dump sugar and fat into the blood stream while sparing protein. This fact partially explains why healthy owners have nice muscles while unhealthy owners do not.

The basic sequence of unhealthy owner's body protein's destruction when they are stressed, between meals or exercising and their blood fuel begins to fall depends on other hormones. The entire group consists of glucagon, cortisol, and epinephrine. Like growth hormone they all facilitate the release of sugar and fat into the blood stream. Unlike growth hormone body protein dismantling becomes fair game when growth hormone levels prove insufficient.

Here again, another vicious process destroys the physiques of countless middle-aged owners until they understand the importance of protecting their body's protein content.

A deeper insight into the abnormal stress response occurring in the unhealthy owner

As if the body structure consumption facilitated by cortisol when growth hormone levels diminish wasn't bad enough, there is an additional obesity perpetuator and muscle loss facilitator to consider. Middle aged and obese owners have an increasing tendency for their adrenal glands to make more cortisol and less DHEA.

Healthy owners secrete two times the DHEA compared to cortisol during the stress response. This fact prevents the catabolic message contained in cortisol from becoming too aggressive. In this way some of the body cells give up structure during the stress response while others continue repairing and rejuvenating.

Around middle age a steep fall off in this optimal ratio between the adrenally released but opposing steroids occurs. Cortisol tends to greatly increase relative to DHEA. Remember, DHEA is the major androgen steroid that instructs the liver to make IGF-1. IGF-1 facilitates muscle nutrition and impedes fat accumulation. A fall off of IGF-1 will necessitate an increased release of insulin. Insulin contains the fat maker message.

This last fact, taken together with the need for increased insulin whenever mental stress occurs, without a compensatory walk off the stress experience, explains how the fat maker message becomes amplified. A falling DHEA secretion rate from the adrenal gland accompanies middle age. A chronic stress response amplifies this problem because more cortisol secretes relative to the DHEA released.

Until these owners improve their DHEA levels, manage their stress and increase their growth hormone promoting activities, these owners' muscles and organs will shrink while their fat increases.

Chapter 2

Mineral Imbalance and Obesity's Creation:
The American Method for Becoming Weak and Fatigued

The mineral content of processed food has been significantly altered from the natural state. "Real food" contains high magnesium and potassium. The body was designed to eat food in the natural state for a cell's electrical charge to remain strong.

Processed food, among other depletions, contains low amounts of these nutrients. The other mineral aberration contained in processed food concerns its unnaturally high sodium content. The food industry adds sodium to prolong the shelf life of these processed products. Shelf life prolongs because bacteria have just as hard a time flourishing with high sodium, as do other cells. Sometimes processed foods contain only relatively low magnesium and potassium content because of all the sodium that has been added. The mineral content of processed food has been significantly diminished from the natural state. Health becomes compromised when an owner eats these processed foods and experiences chronic stress. Stress

44

hormones alter the ability to remove extra salt from the body causing fluid retention. Stress hormones increase the loss of magnesium and potassium from the tissues creating further health consequences.

As a general rule, 4,000 mg of potassium, 1000 mg sodium, 300 mg of magnesium, and 500 mg of calcium are needed every day. Most Americans consume the reverse ratio between sodium and potassium (see mineral table below). The ideal mineral intake ratio applies only when one has normal kidneys and adrenals. Extremes of environmental heat or exercise habits increase sodium requirements. Individual variation occurs on the optimal amounts of minerals needed. This general reference will put most owners back on the path to adequate cellular charge.

When one consumes food that confers these mineral ratios, the body can effectively charge its trillions of cell batteries. This remains true even when an owner experiences stress. If the kidneys have not been damaged from the processed food diet, blood pressure should drop. This will be especially true when a concomitant effort to diminish insulin production and obtain optimal vitamin status occurs.

Seven Reasons the Potassium-Deficient Diet Leads to Old Cells

The importance of ample potassium in the diet has been over looked as a determinant of health. The half-truth about the desirability for a low sodium diet has been substituted for an appropriate emphasis on potassium. There is also a preoccupation about the importance of calcium for the bones while the need for magnesium is largely ignored. The truth would include a balanced discussion of all four important minerals. A real food diet provides the only way that these four minerals correctly proportion in consumption relative to need (see mineral table below). The focus here will be on potassium only to allow clarity on how mineral balance fundamentally prolongs health. Potassium serves as an excellent example for the importance of achieving appropriate ratios of mineral intake.

Optimal potassium:

- Stabilizes and conserves protein
- Protects the kidney
- Enables efficient insulin function
- Maintains the flexibility and health of red blood cells
- Facilitates normal testosterone with normal blood pressure instead of normal blood pressure from medication with a consequent decreased testosterone production
- Increases the electrical charge of the cellular force field
- Promotes optimal cellular function

Potassium and Protein

Potassium content provides a crucial determinant of the cellular ability to hold onto protein. All body cells require adequate potassium content to stabilize their protein content. For every gram of protein, 2.6 milliequivalents of potassium become necessary. Those who consistently eat low potassium diets begin to notice an increase in body fat and a decrease in muscle mass around middle age. Often the loss of muscle largely results from inadequate potassium and/or excess sodium intake. In addition, chronic stress accelerates potassium loss, which contributes to muscle loss as well.

Muscle wasting often results from the body compensating for a potassium deficiency. The body will raid the cells for potassium if it is not provided in the diet. The price of raiding inside the cell storehouses of potassium concerns the loss of cellular protein. When cells give up protein they lose structural and functional integrity. These losses manifest clinically as weakness, fatigue and the look of middle age.

Potassium and the Kidney

It has been known for many years that low potassium diets carry the risk of kidney damage (hypokalemic nephropathy). Stress and a processed food diet further increase the risk of hypokalemic nephropathy. The kidney damage will progress when the potassium deficiency persists. This type of kidney damage can lead to high blood pressure.

A correlation between kidney damage and high blood pressure caused by the potassium deficiency occurs. Many owners in America are not informed about this simple consideration. The longer the time that the owner spends with symptom control, the longer the potassium deficiency will exist and do further damage. The potassium deficiency will not show up in the blood test commonly run at the doctor's office until the total body potassium content severely depletes.

The standard blood test for potassium only measures the two percent tank of body potassium, the blood stream amount (not including the red blood cells amount). The ninety-eight percent tank of body potassium concerns the amount of potassium within the body cells (including the red blood cells). The cell content of potassium will deplete itself to keep the blood stream amount normal until late in the disease process.

The Function and Efficiency of Insulin

The effectiveness of insulin to lower blood sugar requires potassium. Each glucose molecule that enters a cell requires one potassium ion. The more carbohydrates in the diet, the more an owner needs to increase potassium consumption. The more processed the carbohydrates the less potassium content. Only real food contains ample potassium (see mineral table below). Owners who fail to consume sufficient potassium in their diet increase their risk of insulin caused health consequences.

They will become insulin resistant, meaning that for a certain sugar intake one needs more insulin to bring the blood sugar back to normal. The increased insulin required will tend to abnormally stimulate the liver to uptake more sugar. When this occurs the liver will increase its manufacture rate of sticky fat and cholesterol particles (LDL) from the carbohydrate consumed. In other words, the higher blood sugar stimulates the secondary liver sugar removal pathway that does not require potassium. Like other body cells the liver needs a fixed ratio of potassium to sugar in order for glycogen to be made. When potassium becomes deficient the liver then preferentially makes sugar into fat and cholesterol because this process does not require potassium. Only the brain, red blood cells, and liver can take up blood sugar without potassium. All other cells need adequate potassium to feed on the sugar in the blood stream.

Potassium in the cells is sacrificed to the blood stream. Cells donate potassium when more carbohydrates are consumed than there is potassium available. Over time the body begins to be depleted in its cellular potassium. With greater potassium deficiency the ability of the cells to donate potassium for sugar uptake slows. The cells contain less potassium, with chronic potassium deficiency, so they are more reluctant to part with it. Healthy owner's potassium donation occurs quickly to allow the blood sugar to return to normal because they have an optimal supply of potassium relative to sodium in the diet. In contrast, when the potassium deficiency becomes chronic many cells are depleted from optimal potassium content. This process accelerates when an owner finds himself/herself under stress. Stress increases potassium loss.

Many different healing principles must be applied in life. An owner could be practicing other principles of health and fail only because they have not been counseled to turn off the desire of the liver to make more sticky fat. They need to know this important mechanism of how a potassium deficiency will increase insulin's fat maker role in their body.

Red Blood Cell Flexibility

The red cells have the feeblest electrical charges of any cells. These cells often sacrifice their potassium content first to keep the serum potassium level normal. A mild potassium deficiency can be missed when the doctor only checks a serum potassium level.

When red cells lose potassium content their ability to charge their force field diminishes and they weaken. Weak red cells less effectively squeeze out their contents of nutrients, oxygen, and hormones at the capillary level. Red blood cells with diminished flexibility raise blood pressure (see *The Body Heals*, Ferril, 2003).

Testosterone and Blood Pressure

Normal testosterone synthesis with a normal blood pressure is not possible in certain owners that eat a processed food diet. The combination of high salt with low potassium and magnesium commonly lead to high blood pressure. These situations often require treatment in the mainstream medical paradigm with ACE inhibitors. The popular explanation being that ACE inhibitors work because they lower aldosterone production rates. The aldosterone level controls the rate limiting step for testosterone production as well as all other steroids. Conventionally trained physicians are taught to reduce aldosterone levels. The fundamental role aldosterone plays in all the steroids biosynthesis is largely down played or ignored.

A real food diet allows for normal aldosterone levels and normal blood pressure. It contains high potassium and magnesium with low sodium. In some owners, only when these minerals become properly proportioned can blood pressure problems be avoided. The body was designed to consume the mineral proportions contained in the real food diet. Conversely, the processed food diet violates the basic body design theme in that the mineral proportions are drastically altered. Individual variation occurs for how long a body can tolerate an aberrant mineral intake.

Correcting chronic reversed minerals intake and eventual total body potassium depletion may take up to six months before the blood pressure drops maximally. Just as it takes a long time to deplete body mineral composition, it takes a long time to replenish minerals.

The analogy to understand how a processed food diet will eventually harm owner's cells is similar to purposely violating the mineral composition of battery fluid. Like car batteries, human batteries were designed for specific mineral intake. The body possesses remarkable resilience to chronically ingesting reversed mineral consumption, but a tolerance limit exists. Cells that reach their tolerance limit behave in electrically aberrant ways. Electrical aberrancy proves the culprit, but aldosterone takes the blame.

Western-trained physicians are taught to decrease aldosterone levels when blood pressure elevates. Aldosterone will only elevate blood pressure when the minerals within are imbalanced. The elevated aldosterone effect further exacerbates blood pressure when stress also occurs because it increases potassium and magnesium losses while retaining sodium. With sodium retention comes fluid retention. The fluid retained because of stress and reversed minerals intake causes the blood pressure to rise.

Owners who are stressed, but consume the proper mineral ratio become less vulnerable to fluid retention induced high blood pressure. They will tend to tolerate a higher aldosterone level that allows a proper stimulus for their body's production of other steroids. Proper mineral consumption satisfies one determinant of the of the steroid synthesis rate.

Most physicians are schooled in the vague principle of a low sodium diet only. Very few doctors are aware of food types that contain low sodium (such as real food found in the mineral table below). Even fewer physicians understand the importance of increasing the potassium and magnesium containing foods in the diet. Real food contains high magnesium and potassium (see mineral table below). Processed food has less magnesium and potassium content. Processed food also has relatively large amounts of sodium added to it, which worsens the mineral ratios imbalance further. Only when physicians understand the facts about the different natural mineral proportions can they begin to counsel on diets that will heal. Owners heal from correcting cellular charge related health problems.

The Cellular Force Field

The amount of potassium contained within a cell forms a central determinant for the cell's ability to keep out unwanted molecules. The proper potassium level permits the membrane to maintain a protective electrical charge. When a potassium deficiency exists, less protection occurs. This leads to the destruction of the cell. The chronic consumption of reversed mineral proportions contained in processed food diets leads eventually to decreased cellular potassium levels. Sodium is over abundant in America's food supply and potassium deficiency is prevalent. To maintain cellular health, there must be proper proportions between sodium and potassium within the body.

Potassium Availability and Cell Charge

Cell charge is directly proportional to the amount of potassium available in the cell. Optimal cellular function requires adequate potassium content within a cell. Owners with diminished potassium content have weak cells that cannot accomplish the same function as properly nourished cells. The availability and proportions of potassium, sodium, calcium, and magnesium determine how much work a cell can perform.

Owners that understand the importance of this mineral ratio take the time to create a diet that reflects this. A real food diet will naturally provide these ratios and amounts (see real food list in torture chamber diet section). When these considerations are incorporated into an owner's life, another healing principle is better satisfied.

Mineral content of some common foods, in milligrams

Breads, Rolls, Etc	Amount	sodium	potass.	calories	mag.	calcium
White Bread	1 slice	142	29	76	0	32
Rye Bread	1 slice	139	36	61	0	20
Whole Wheat Bread	1 slice	132	68	61	0	20
Biscuit	1 (2"dia.)	185	18	104	0	0
Cornbread	2 1/2 sq.	263	61	178	0	133
Pancake	1 (6" dia.)	412	112	164	16	60
Waffle	1 (7" dia.)	515	146	206	13	143
Graham Cracker	2(2 1/2" sq.)	95	55	55	14	12

		sodium	potass.	calories	mag.	calcium
Brown Rice	1 c + salt	550	137	236	86	20
White Rice	1 c + salt	767	57	223	26	24
Bran Flakes	1 cup	207	137	106	108	26
Corn Flakes	1 cup	251	30	92	16	6
Oatmeal	1 c (cooked)	523	146	132	57	22.5
Pufffed Rice	1 cup	148	33	140	0	10
Wheat Flakes	1 cup	310	81	106	108	100
Wheat Flour	1 cup	130	0	499	0	0
*Egg Noodles	1 cup	3	70	200	31	19
*Macaroni	1 cup	1	103	192	20	8
*Spaghetti	1 cup	1	103	192	0	0

Beverages	Amount	sodium	potass.	calories	mag.	calcium
Coffee, Instant	1 Tblspn	3	87	3	*80	*50
Coffee, Regular	1 cup	2	65	2	10	3
Beer	12 oz	25	90	150	46	36
Gin, Rum, Vodka	1 oz (80 proof)	0	1	65	0	0

Sweets	Amount	sodium	potass.	calories	mag.	calcium
Angel Food Cake	1/6 cake	340	106	322	0	88
Brownie	small	50	38	97	0	0
Chocolate Bittersweet	1 oz	1	174	135	30	6
Chocolate, cupcake	1 piece	74	35	92	0	0
Chocolate chip cookies	10 (2.5" dia.)	421	141	495	0	32
Chocolate Syrup	1 oz	20	106	92	24	3
Gelatin, sweet	3 oz	270	0	315	0	1

Glandular Failure-Caused Obesity

Honey	1 Tblspn	1	11	64	0	1
Jelly	1 Tblspn	3	14	49	trace	4
Sherbert, Orange	1 cup	19	42	259	16	104
Sponge Cake	1/6 cake	220	114	196	0	42
Sugar, Brown	1 cup	44	499	541	0	187
Sugar, White	1 cup	2	6	770	trace	trace
Sugar, powdered	1 cup	1	4	462	trace	0

Fruits	Amount	sodium	potass.	calories	mag.	calcium
Apple	1 (2 1/2" dia.)	1	116	61	6	10
Apricots, Fresh	3 medium	1	301	55	8	15
Apricots, Dried	5 lg halves	6	235	62	11	10
Banana	1 medium	1	440	101	33	7
Blackberries	1cup	1	245	84	28	46
Cantalope	1/2 (5" dia.)	33	682	82	28	28
Cherries, Sweet	10 count	1	129	47	8	10
Dates	10 count	1	518	219	29	27
Figs	1 piece	1	126	52	8	18
Grapefruit	half	1	132	40	10	13
Grapes	1 cup	3	160	70	3	9
Honeydew Melon	half (6 1/2" dia.)	90	1881	247	9	8
Orange	1 medium	1	290	66	15	56
Peach	1 (2 3/4" dia.)	2	308	58	6	5
Pear	1 (2 1/2" dia.)	3	213	100	5	10
Pineapple	1 cup	2	226	81	22	12
Plum	1 (1"dia.)	0	30	7	<1	1
Prune, Dried	10 medium	5	448	164	51	58
Raisins	1 tablespoon	2	69	26	3	5

Raspberries	1 cup	1	267	98	22	28
Strawberries	1 cup	1	244	55	16	22
Tangerine	1 (2 3/8" dia.)	2	108	39	10	12
Watermelon	1 cup	2	160	42	18	14
Avocado	1 medium	21	1097	324	70	19

Fresh Vegetables	Amount	sodium	potass.	calories	mag.	calcium
Asparagus	1 cup	3	375	35	22	30
Beans, Lima	1 cup	3	1008	191	126	54
Beets	1 cup	81	452	58	28	22
Broccoli	1 cup	34	868	72	38	72
Carrot	1 medium	34	246	30	11	19
Celery	1 stalk	50	136	7	4	16
Corn, Sweet (no butter, no salt)	1 ear	0	131	70	34	2
Cucumber	1 large	18	481	45	33	42
Eggplant	1 cup	2	300	38	10	30
Lettuce, iceberg	1 head (6" dia.)	48	943	70	48	102
Onion	1 cup	17	267	65	16	32
Peas	1 cup	3	458	122	34	62
Potato, Baked	1 medium	6	782	145	55	20
Potato, Boiled	1 medium	4	556	104	30	7
Radishes	10 large	15	261	14	4	9
Spinach	1 cup	39	259	14	158	244
Sweet Potato	1 medium	15	367	272	32	70
Tomato	1 medium	4	300	27	13	6
Watercress	1 cup	18	99	7	8	40

Glandular Failure-Caused Obesity

Canned Vegetables	Amount	sodium	potass.	calories	mag.	calcium
Asparagus	14 1/2 oz can	970	682	74	22	34
Beans, Green	8 oz can	536	216	41	18	36
Beans, Lima	8 oz can	1070	1007	322	94	50
Beets	8 oz can	535	379	77	40	34
Carrots	8 oz can	535	272	64	22	62
Corn, Creamed	8 oz can	585	241	203	44	8
Peas	8 oz can	569	231	159	22	44
Spinach	8 oz can	519	550	42	132	144
*Tomatoes	8 oz can					

Dairy Products	Amount	sodium	potas.	calories	mag	calcium
American Cheese	1 oz.	322	23	105	6	124
Blue Cheese Dressing	1 Tblspn	164	6	76	0	0
Cheddar Cheese	1 oz.	147	17	84	8	204
Cream Cheese	1 oz.	80	25	110	2	23
Cottage Cheese	1 cup	580	144	172	14	154
Parmesan Cheese	1 oz.	208	42	111	14	390
Swiss Cheese	1 oz.	70	29	105	10	272
Butter (salted)	1 stick	1119	26	812	2	27
Butter (unsalted)	1 stick	<1	<1	812	2	27
Buttermilk (cultured)	1 cup	319	343	*88	27	285
Skim Milk	1 cup	127	355	*88	28	302
Whole Milk	1 cup	122	351	159	33	291
Evaporated Milk	1 cup	297	764	345	60	658
Heavy Cream	1 Tblspn	5	13	53	1	10

Ice Cream (no salt)	1 cup	84	241	257	9	88
Hot Chocolate	1 cup	120	370	238	24	93
Hot Cocoa	1 cup	128	363	243	0	0
Egg Yolk	1 medium	8	15	52	1	23
Egg white	1 medium	42	40	15	4	2
Egg Broiled	1 medium	54	57	72	5	25
Yogurt, Plain	1 cup	115	323	152	26	274

Meat and Poultry	Amount	sodium	potas.	calories	mag.	calcium
(Beef)						
Corned Beef Hash	1 cup	1188	440	398	14	9
Frankfurter	1 med.	627	125	176	6	7
Heart	1 oz	29	66	53	0	0
Hamburger	2.9 oz	49	221	235	5	2
Liver	3 oz	156	323	195	0	0
Rib Roast	6-9 ribs	149	680	1342	22	11
Flank Steak	3 oz	45	207	167	26	7
Porterhouse Steak	11 oz	155	680	1400	28	9
Sirloin Steak	11 oz	173	793	1192	32	12
T-Bone Steak	11 oz	152	660	1431	28	9
(Lamb)						
Chop	1 med.	51	234	341	27	28
Roast	3 oz	60	273	158	36	16

(Pork)

Glandular Failure-Caused Obesity

Bacon	1 slice	123	29	72	<2	<1
Chops	3 oz	47	214	300	15	6
Ham, baked	3 oz	770	241	159	18	5
Roast	3 oz	698	218	281	0	0
Spareribs	2 pieces	65	299	792	0	0
(Veal)						
Loin cut	3 oz	60	570	220	9	10
Roast	3 oz	57	259	229	21	23
(Chicken)						
Broiled	4 oz	75	310	154	28	19
*Light Meat	4 oz	60	240	120	25	18
*Dark Meat	4 oz	45	100	180	27	19
(Turkey)						
White Meat	4 oz	70	349	150	29	29
Dark Meat	4 oz	42	169	87	26	26

Fresh, Fish & Seafood		sodium	Potass.	calories	Mag.	calcium
Bass, striped	3 oz	0	0	168	69	27
Clams	4 clams	144	218	56	52	12
Cod	3 oz	93	345	144	16	36
Crab	1 cup	0	0	144	78	0
Flounder	3 oz	201	498	171	15	27
Haddock	3 oz	150	297	141	27	33
Halibut	3 oz	114	447	144	37	72
Lobster	1 cup	305	261	138	138	80
Mackerel	3 oz	0	0	201	9	63

Glandular Failure-Caused Obesity

Oysters	3 small	21	34	19	39	45
Salmon	3 oz	99	378	156	180	29
Shrimp	3 oz	159	195	192	45	30

Chapter 3

Exercise and Obesity:

Big Muscles or Lots of Fat

Muscle size and strength provide tangible evidence for the power contained in hormone quality and proper nutrition. Directly observing these two important determiners of muscle and ligament function reveals that something very important often turns up missing from the physical exam in America today.

The science exists which explains the importance of optimal hormone (types and amounts) and nutritional building blocks for muscle tissue. All other performance enhancers marginally affect muscle size without these primary determinants. The failure to include this in one's annual evaluation, in the clinical setting, often contributes to an unnecessary acceleration into old age. Obesity describes only one component of the aging process.

The first determiner, the quality of hormones, involves the fact that muscle cells only perform as they are directed. The most powerful directors concern the level one hormones. They instruct a muscle cell's DNA program (genes) activity level.

The types and amounts of these most powerful hormones determine which genes turn off and which genes turn on.

The steroids, vitamin A, and thyroid hormones comprise the only hormones that contain this direct ability. The quality and amounts of the different steroids determine whether the message to the muscle cell DNA provides coherence. Coherence or incoherence to energy direction is analogous to a cellular 'melody' or cellular 'noise.' The incoherent message of noise leads to old age. One way to avoid old age depends on which DNA programs activate or suppress. The quality of the message content determines where in the continuum the owner lies between chaotic genetic programs all the way up to the message melody of health. The DNA activity powerfully determines how big and strong a muscle cell becomes.

Muscle cells provide excellent examples of the importance for balance between anabolic and catabolic message content. Only when proper balance occurs between these two opposing messages can the DNA program direct buildup activities appropriately within the muscle cell. Extremely high anabolic message content occurs in the body builder. Large muscles only become possible with high androgen message content. High androgen message content directs the muscle cell DNA to increase cellular infrastructure investment activities. Muscle development provides an example of the central role of anabolic steroids. Anabolic steroids direct the DNA program to increase cell build up and repair activities.

The muscle cell response to steroids is no different than the other cells throughout the body. Other cells in the body respond to appropriate amounts and timing of their preferred anabolic steroid. The difference between muscle cells and other body cells concerns the fact that each body cell type has its preferred anabolic steroid. For example DHEA is the preferred steroid within the brain. Whatever the serum DHEA level, the brain concentrates it by a factor of five to six. DHT is the preferred steroid for nice skin. The muscles prefer testosterone for their maximal development message.

Muscle cell message content needs to be counterbalanced by the catabolic steroid, cortisol. Appropriate but not excessive cortisol message content maximizes energy while an owner exercises. The cortisol message also prevents excessive soreness following exercise (an additional message content of cortisol called an anti-inflammatory effect).

Remember that the adequate release of growth hormone moderates the catabolic protein combusting message of cortisol. If growth hormone levels prove inadequate when an owner exercises excessive protein structure will dismantle. Healthy bodies contain appropriate amounts of both cortisol and growth hormone while their muscles exercise. Balance between opposing hormones fundamentally prolongs youthfulness. The hormones giveth and the hormones taketh away. In other words, in the youthful state they giveth and in the unhealthful state, the imbalanced message (noise) taketh away.

Muscles need adequate stimulus (exercise) before the gonads and adrenals respond and produce increased message content from the anabolic steroids. A relationship between muscle use and steroid message content exists. Maximal muscle development depends on both processes. Failure on either end of this equation leads to little muscles.

Exercise results in soreness and annoying injuries (strains) when adrenal and gonad function impair. Any owner who suddenly begins to become excessively sore following modest workouts needs a steroid hormone evaluation. Replacement with real steroids may be indicated when a severe hormone deficiency reveals itself. Many out of shape owners who get in shape respond nicely to improved diets and nutrients alone (see below).

Once a complete inquiry into the muscle hormone status of the level one-hormone types occurs, the 'lesser' hormones can be considered (see appendix A on the hierarchy of hormones). An example of a level 2-hormone type is insulin-like-growth-factor (IGF-1). IGF-1 occurs at amounts over 100 times more plentiful in the blood stream of healthy owners than those of insulin. IGF-1 helps muscle cells obtain their nutrition within the blood steam. IGF-1 facilitates the anabolic steroids desire to build bigger and stronger muscles. When IGF-1 within the blood stream lowers, the anabolic steroid message has less ability to build cells like muscle cells. Muscle cells cannot build themselves up without the proper nutritional molecular building blocks (see digestion section for specifics). IGF-1 and insulin facilitate the absorption of nutrition by muscle cells. Insulin needs increase when IGF-1 falls because insulin becomes the backup hormone (fuel nozzle) for muscle cell fuel intake.

Physician's Sidebar

The relative roles of IGF-1 and insulin in muscle cell nutritional needs have been largely ignored within the mainstream medical approach. Even though, in the healthy owner, IGF-1 occurs at 100 times the normal insulin levels, the increase of insulin receives the advertising dollar. The trouble with this approach involves the fact that insulin produces many side effects, which IGF-1 does not share. Examples of the differences between the insulin message content compared to IGF-1 within the body are: insulin causes increased cholesterol and triglyceride synthesis within the liver, insulin more rapidly degrades without adequate IGF-1, increased insulin increases appetite, insulin rises after eating while IGF-1 rises during fasting.

Insulin and IGF-1 facilitate the uptake of muscle fuel. However, in healthy owners, insulin facilitates a relatively small amount of fuel uptake within muscle cells because it occurs in much smaller amounts (less than one percent of IGF-1 levels). Additionally, insulin only elevates maximally following a carbohydrate meal. IGF-1 levels in healthy owners tend to stay up between meals. IGF-1 has a half-life of four days while insulin's half-life is less than ten minutes. In addition, IGF-1 levels rise slightly between meals (fasting). The simultaneous rise in IGF-1 and the release of sugar and fat into the blood stream, which growth hormone commands, provides fuel to healthy muscles between meals.

Healthy muscles depend on adequate IGF-1 levels for their nutritional needs between meals. They also depend on adequate insulin following meals, which directs sufficient storage of sugar and fat for the next between meal states. Between meals growth hormone releases when the brain senses a fall in blood sugar. The growth hormone released causes the stored sugar, fat and IGF-1 to release. Sufficient IGF-1 between meals allows the muscles to receive nutrition without the need for insulin. Healthy people have no need for insulin in the fasting state.

Syndrome X individuals, because their IGF-1 has fallen, need insulin in exaggerated amounts between meals (insulin resistance). The liver always receives the highest concentration of pancreatic hormonal secretions. Liver and fat cells also have the highest amount of insulin receptors occurring at 200,000 per cell. The increased insulin output stimulates the liver and fat making machinery. Syndrome X owners, then, have conflicting message content within their livers during the fasting state.

The first conflicting aberration concerns their diminished growth hormone out put while between meals. Increased epinephrine, glucagon and cortisol then need to release in order to maintain the blood sugar level (fuel level) between meals. However, the fall in growth hormone has two health consequences. One is that body protein stores become fair game for dismantling when growth hormone levels fall off. Second involves the fall off of IGF-1 release. This last fact sets up the overall second conflicting message at the level of the liver. Simplistically the conflict involves the competing message between the increased insulin needed to shore up fallen IGF-1, which also tells the liver to store fuel and the simultaneous release of the counter hormones, cortisol, glucagon and epinephrine, telling the liver to release fuel. Until someone helps these owners to realign their hormones that instruct their liver, these owners will continue on the accelerated path to an old body.

Popular strategies abound that purport to raise growth hormone secretion rates. All of them ignore the fact for why the pituitary secretes growth hormone in the first place. Growth hormone secretes to protect body protein content yet keep the blood fuel level adequate between meals or when exercising.

Cognizance of this fact allows one to see that the popular approaches have their effect by peripheral pathways.

Certain amino acids are fastidiously protected within the body from fuel combustion. Popular strategies raise growth hormone levels in some owners because these same amino acids when taken as supplements set off the pituitary alarm that important amino acid blood levels are elevated. Examples of important amino acids that the body fastidiously protects are: arginine, ornithine, glutamine and lysine. But what about those owners who have a pituitary defect acquired around middle age that no longer allows sufficient growth hormone release?

Many Syndrome X owners have such a defect. They are doomed to die prematurely unless they receive counsel on how to raise their growth hormone levels. Here again popular growth hormone replacement protocols fail to inquire about liver health and androgen levels before prescribing growth hormone replacement injections. Hence, all the media attention to the supposed risk for growth hormone induced diabetes. Healthy livers that receive adequate androgen message content will curtail a diabetes tendency with growth hormone treatments (see liver chapter).

IGF-1 levels depend on two basic factors. First, concerns the hormonal factor. The liver needs adequate direction from thyroid, DHEA or testosterone and cortisol. When all three of these hormone types occur at normal levels, these level one hormones instruct the liver DNA to manufacture IGF-1 hormone and its binding proteins properly. However, number two, the life style factor, allows the release of IGF-1 from the liver. The lifestyle factor controls growth hormone [GH] release rates. Growth hormone releases when the blood fuel level decreases. Common situations for which the blood fuel falls are: exercise, fasting and between meals. Increased estrogen levels inhibit the release of IGF-1 by GH (see estrogen section). Owners who have healthy levels of IGF-1 also have healthy levels of Thyroid, DHEA, testosterone, cortisol, estrogen and GH.

The hormone report card evaluates these when health diminishes. Paradoxically, mainstream medicine does not routinely inquire about these important hormone

considerations. **The suspicion is that mainstream doctors are not taught to think about these hormone relationships. I wasn't encouraged. The science is all there, although it presents in a convoluted and disjointed fashion. As long as there continues to be divergent and multiple ways to say these important facts physicians will remain in the dark. Be kind to your physician and help him/her to learn anew.**

Maintaining balance between IGF-1 and insulin form an important biochemical determinant of youthfulness within the muscles. It is also important to avoid the popular practice of receiving growth hormone injections without a proper evaluation of the IGF-1 levels. Without a proper evaluation of one's livers ability to increase IGF-1 levels, GH will tend to promote high blood sugars and therefore increase insulin output. Increased insulin output associates with all the negative effects on body physique discussed in the insulin subsection.

This last fact explains the negative publicity associating growth hormone injections with breast and abdominal fat growth. Normal livers respond to growth hormone by releasing stored IGF-1 along with sugar and fat. The released IGF-1 delivers the sugar and fat released to the organs and muscles thus negating the need for insulin in the fasting state. Unhealthy livers can still release sugar and fat but release diminished amounts of IGF-1 and insulin release rates consequently need to increase to pick up the slack in the elevated blood sugar. Higher insulin levels always go to the liver first and thus stimulate the fat and cholesterol making liver machinery. The extra fat and cholesterol deposits within the breast and abdomen. All these side effects could be avoided if someone first helped the liver to heal (see liver chapter).

The level three hormones (see hierarchy of hormones) affect muscle performance by opening up or closing down the blood supply within the exercising muscle. Deficiency of the level three hormones leads to diminished muscle performance. Adequate supplies of the level three hormones depend on adequate protein meals and many vitamins. Examples of level three hormones are: epinephrine, nor-epinephrine,

Physician's Sidebar

High blood sugar stimulates insulin increases but low blood sugar stimulates IGF-1 release along with sugar and fat. The opposite is also true. High blood sugar eventually decreases IGF-1 levels but low blood sugar decreases insulin release. When IGF-1 releases there has been a preceding release of GH. The GH release response results from a falling blood sugar, which causes stored sugar to be dumped by the liver into the blood stream. The simultaneously released IGF-1 then causes the liver released sugar to be used. IGF-1 levels maintain body cell fuel levels between meals. Insulin maintains liver storage levels of sugar and fat following meals. In this way healthy muscles have access to fuel at all times.

histamine, serotonin, and dopamine. All of these hormones have short active life spans of several minutes. Therefore, the amount of the level three hormones constantly determines the tone of the blood vessels every few minutes.

The useful analogy here regards the water system network that underlies many large cities. Fluctuations in demand because of time of day and location necessitate that water engineers open up or tighten down available water supply with numerous different and strategically located check valves throughout the city water supply system. Similar processes operate within the body. Understanding ways to direct maximal blood supply into the performing metabolically hungry areas of the body confers a performance advantage on its owner. Under optimal conditions, the body reroutes a tremendous increase in blood flow to the active area. Shunting blood away from the less active areas enhances the effect.

The adrenal gland contains two layers. The outer layer makes up the cortex. The adrenal medulla describes the inner layer and manufactures some of the level three hormones.

In order to understand muscle performance the inner core (the adrenal medulla) of the adrenal gland's hormonal products needs to be understood. Here powerful hormones release and prevent unconsciousness from occurring when one stands up. The arterial muscle layer contracts when one stands. The arterial muscles will contract when enough level three hormones instruct them to do so. The ability for the arteries to contract or relax is determined in part by the adrenal medulla. The other part of contraction versus relaxation, in the blood vessels, concerns the autonomic nervous system. Here the discussion focuses on the hormone component of blood vessel caliber because nutrition powerfully affects this ability (see below).

The adrenal medulla is largely responsible for performing the task of the body's 'water engineers'. The city planners design the water system underneath the city but the water engineers constantly decide which valves to turn up and which to turn down in flow. Incompetent water engineers on the staff create inefficient water delivery to certain areas of the city. This analogy simplifies what the adrenal medulla does in regards to directing blood flow within the vessels.

The level three hormones task of directing blood flow requires certain molecular parts availability. Epinephrine proves the most desirable of the level three hormones in exercising muscle. Exercising muscle needs sufficient epinephrine levels to direct certain blood vessels to open. When the blood vessels supplying muscle open an increased supply of fuel and oxygen delivery occurs. Some of the other level three hormones, if present, compete with this process (see below).

The second determinant of powerful muscles, alluded to above, concerns the nutritional status of the body. Epinephrine manufacture requires specific nutrients. The adrenal gland needs an adequate supply of either phenylalanine or tyrosine and also numerous cofactors (vitamins). If one or more cofactors become deficient then the adrenals cannot manufacture the most beneficial types of exercise performance enhancing hormone within this class, epinephrine. The necessary cofactors needed for the synthesis of epinephrine are: tetrahydrobiopterin (made from folate), vitamin C, vitamin B6, and SAMe. Since SAMe deactivates within the body at the rate of one billion times a second, it needs to be recharged with the following nutrients: vitamin B12, folate, serine, and methionine. (*Cooney*)

Epinephrine is a member of the bioactive amines, listed above. Scientists call the bioactive amines either hormones or neural transmitters. Their site of action determines which class they are in. When bioactive amines are between two nerve endings, the synapse, scientists call them neural transmitters. If they discharge into the blood stream (the adrenal medulla is a major site for this but there are other sites through out the body) scientists call them hormones.

When bioactive amines course in the blood stream they act as the water engineers, as discussed above. All water engineers are not created equal. Peak performance athletes predictably receive only the best 'water engineers' directing their blood flow. Most owners fail to understand ways to enhance the level three-hormone mixture to obtain more competent water engineers. More competent water engineers lead to better blood flow to the exercising muscles. Nutritional deficiencies prevent the manufacture of adequate epinephrine, which leads to a prematurely diminished athletic performance.

For maximum performance abilities within the muscles epinephrine can be thought of as the master water engineer. It contains the informational content to open up the blood vessel diameter leading to the exercising muscles, liver and the heart. Increasing blood flow within the areas of increased metabolic demand obviously allows for better delivery of oxygen, nutrients and the increased ability for waste removal (carbon dioxide, lactic acid, spent cofactors, etc.). The liver needs an increased blood supply during exercise because it removes and reprocesses the huge amounts of lactic acid generated within exercising muscles. Also remember the liver performs as the main fuel releasing organ in the body (see liver chapter).

Epinephrine requires all the above cofactors availability because its biosynthesis lies at the end of the assembly line for the bioactive amines manufacturing process. When nutritional deficiencies cause the synthesis of epinephrine to decrease there becomes less instruction for the blood vessels to open up during exercise. Normally during exercise the blood supply to the heart, skeletal muscles and liver increases because of epinephrine's presence. However, when epinephrine becomes deficient nor-epinephrine levels will rise. The trouble with increased levels of nor-epinephrine concerns the fact that its message tightens down the blood supply to the heart, skeletal muscle and liver. The health consequence here results from less fuel and oxygen delivery to these metabolically active tissues during exercise. Also, the nasty side affect of abnormal blood pressure elevation occurs (explained in Appendix B).

When optimum conditions prevail (youthfulness) the adrenal medulla will make 90% epinephrine. Under healthy conditions (to be reviewed shortly) the adrenal only makes 10% nor-epinephrine and dopamine. This has important implications for those owners who desire to feel as good as good as possible when they exercise.

One of the most common nutritional deficiencies within the adrenal medulla occurs in what is known as the **methyl donor group of substances**. When a methyl donor deficiency occurs within the adrenal medulla, there arises the inability to convert nor-epinephrine into epinephrine. When these molecular part deficient adrenals secrete, during exercise, their ability to dump epinephrine into the blood stream diminishes. Epinephrine release proves inadequate because of the methyl donor deficiency (see methyl donor system appendix B). Only epinephrine contains the message to direct blood flow increases to exercising muscles, the heart, and the liver. Exercise performance critically depends on the epinephrine message content to increase blood supply to these areas. Unless epinephrine releases the exercising owner will suffer decreased blood supply to his heart, liver and muscles. In addition, his blood pressure will rise abnormally high because only epinephrine will moderate blood pressure.

Many owners take the expensive SAMe in a pill to recharge their methyl donor status. This is not always necessary or best. Methyl deficient owners degrade SAMe to homocysteine. Elevated homocysteine levels therefore provide an excellent marker for those owners who suffer this deficiency. Supplementing with adequate serine, vitamin B12, vitamin B6, and folate often will recharge the deficient methyl donor state. Remember that these bioactive amines within the adrenal derive from tyrosine or phenylalanine. A functional digestive tract needs to be involved (see digestion section). There are other cofactors needed to manufacture epinephrine from tyrosine: tetrahydrobiopterin (made from folate), vitamin C, vitamin B12 and adequate cellular magnesium.

Chapter 4

More Glandular Secretions That Affect Body Weight

One of the Pituitary Glandular Secretions: Growth Hormone

The most misunderstood and neglected role of the counter response hormones to insulin is growth hormone (GH). This neglect occurs because its name leads owners down an erroneous mental image path. A major effect of growth hormone concerns its ability to cause the liver to release its stored IGF-1. Growth hormone also conserves the protein content of the muscles and organs, while in the fasting state or while exercising.

Physician's Sidebar

Growth hormone's protein conservation message content explains where its name originates. Protein conservation proves as a pre-requisite for growth to occur. Growth hormone, other than its stimulatory effect on cartilage cell growth, has few direct effects on the tissues. One additional direct affect concerns its ability to act like the other three counter hormones to the message of insulin, at the level of the liver, with two important exceptions. First, involves the fact that unlike epinephrine, cortisol and glucagon, growth hormone has a powerful protein sparing effects. Growth hormone's message content in the liver inhibits the conversion of amino acids into sugar (a protein sparing effect). The second difference from the other insulin counter hormones involves the fact that GH directs the liver to release a special hormone called insulin-like growth factor type 1 (IGF-1). Like the other counter hormones to insulin, it stimulates the release of sugar stored as liver glycogen into the blood stream. Also like the other counter hormones to insulin it stimulates the liver to release stored fats into the blood stream for fuel, as well.

Insulin-like growth factor type 1 (IGF-1) can only release from the liver with the direction of growth hormone's presence. Mainstream medical confusion arises from the fact that the affects of IGF-1 message content directly oppose the initial fuel release effects of growth hormone. IGF-1 release occurs simultaneously to the growth hormone directed liver release of sugar and fat into the blood stream. If one sees the overall effect in the sequential release of growth hormone followed by IGF-1, this begins to make more sense. The more IGF-1 released, the less insulin needed. The less insulin needed the less fat maker message within the body.

Growth hormone stimulates the release of fat and carbohydrates form liver stores into the blood stream. The second part of its message involves the simultaneous release, from the liver, of adequate insulin-like growth factor (IGF-1) into the blood stream also. The IGF-1 hormone in the peripheral tissues (blood stream) behaves very much like insulin does in the liver and body fat cells. The body needs less insulin when the liver secretes adequate IGF-1. When adequate growth hormone has stimulated sufficient IGF-1 release into the blood stream, the peripheral tissues, like muscle, are facilitated to procure fuel (carbohydrate and fat). This describes the insulin-like effect of IGF-1. However, remember that IGF-1 rises between meals, which feeds the cells in the fasting state. In contrast, insulin rises following meals, which allows the liver to store sufficient fuel to allow it to supply a constant blood fuel until the next meal.

Sufficient release of IGF-1 negates the initial increase in blood sugar and blood fat caused by the presence of growth hormone's message to the liver. Mechanistically IGF-1 behaves like insulin in the peripheral tissues. IGF-1's presence instructs the peripheral tissues to take up the fuel released by growth hormones presence. Many clinicians fail to appreciate this sequential arrangement that operates in the healthy population. Both IGF-1 and insulin bind to some of the same cell receptors. This makes sense when one realizes their similar message content regards instruction of different cells within the body to take up fuel out of the blood stream.

Different cell types have different affinities for IGF-1 and insulin and different cell receptor concentrations for either insulin or IGF-1. For example, the liver and fat cells have the highest amount of pure insulin-type receptors of any other cell type. There are about 200,000 insulin receptors per fat or liver cell. Insulin directs these tissues to store fuel. Here the IGF-1 does not bind well to these types of insulin receptors and therefore has essentially no effect in the creation or maintenance of body fat. In contrast, the IGF-1 receptors are found throughout most of the rest of the cells in the body. IGF-1 blood levels, in healthy owners, occur at levels 100 times that of insulin levels. This makes sense since there are roughly one hundred times the cells that prefer IGF-1 to insulin.

Many disease processes have their origins in a falling IGF-1 level. When the IGF-1 level falls, insulin needs to be secreted in abnormal amounts. Increased insulin has health consequences that increased IGF-1 does not share.

A major advantage of adequate IGF-1 to that of higher insulin involves the fact that insulin levels determine the amount of body fat. Fat is the major stored fuel type because the body possesses a limited ability to store sugar as glycogen. Total storage capacity for glycogen is about 500 grams (about 2200 calories). About four hundred grams store in the liver and the other one hundred grams store in the muscles. Sedentary and well-fed owners have little opportunity to draw down these stored forms of sugar. The more sedentary and well fed the owner, the more carbohydrate that will be channeled into their liver's making fat and cholesterol. **The body is smart and consistent.** Insulin directed pathways are designed with fuel storage in mind. The major fuel storage sites occur in arterial macrophages, the liver and fat cells.

Certain genetically predisposed owners have a higher insulin secretion on a daily basis on a similar diet as compared to normal owners. The increased insulin responding owner means these owners create more message content that directs their livers to make carbohydrates into LDL cholesterol. In these same owners, practices that increase growth hormone will result in increased IGF-1 release from the healthy liver. Increased levels of IGF-1 share essentially none of the liver stimulation effects that lead to increased LDL cholesterol manufacture. This will help lessen the need for insulin (the carrier of the fat and cholesterol maker message) secretion by facilitating the cells to uptake sugar from the blood stream, beyond the liver.

Healthy owners use most of their insulin production at the level of their liver following a meal. This situation allows low insulin needs because these owners produce sufficient growth hormone between meals. Insulin levels rise whenever: an owner consumes carbohydrate, they chronically consume a potassium-depleted diet, experience unmanaged mental stress, and/or when their IGF-1 levels fall off. Less insulin need occurs when IGF-1 levels increase because it facilitates the removal of sugar from the blood stream. Unlike insulin, which has a major effect on the liver and fat cell's ability to remove sugar out of the blood stream, IGF-1 has its effect in the periphery cells (muscles and organs). IGF-1 competes with insulin as to where the extra nutrition sucks out of the blood stream. Higher IGF-1 favors increased nutrition procurement for the muscle and organs cells. Higher insulin levels favor the uptake of nutrition out of the blood stream by liver and fat cells.

There are two major stimuli for growth hormone's release. The GH releasing stimulus results from either fasting or intense exercise. Both of these conditions produce a decrease in blood fuel levels. IGF-1 releases, along with sugar and fat, from a normal liver after growth hormone levels rise. The overall scheme involves the initiator, low blood fuel. This decrease causes an initial rise in growth hormone that initially directs the liver to release IGF-1, sugar, and fat into the blood stream. The released IGF-1 acts like peripheral insulin in facilitating the organ and muscles uptake of the released fuel from the liver. In this way, the body has a mechanism for ensuring that appropriate amounts of fuel are in the blood stream between meals and when physical exertion draws down the blood fuel level.

Exercise has a powerful contributory effect on the amount of growth hormone released and hence, IGF-1 levels. This release does much the same thing in the periphery that insulin does in the liver and fat storage cells. However, low blood sugar effects are prevented because growth hormone also directs the liver to dump sugar into the blood stream while the liver also releases IGF-1. The design of IGF-1 facilitates the peripheral cells uptake of fuel out of the blood stream that, in these cases, growth hormone started. Where growth hormone production falls off and hence IGF-1 as well, the need for insulin production increases. In these unhealthy situations, insulin must pick up the slack in the periphery (muscles and organs). This process describes one of the mechanisms for insulin resistance (liver chapter).

Increased IGF-1 levels occur for the opposite reasons of increased insulin levels. The increase in IGF-1 in the exercising or fasting state facilitate cellular uptake, of the growth hormone stimulated liver release, of sugar and fat into the blood stream. However, the fuel storage adequacy in the liver depends on enough insulin directing the liver to suck up nutrition for storage purposes following a meal. Without sufficient insulin, there would be no stored fuel to release when GH directed the release of fuel and IGF-1 from the liver. In this way, the healthy body balances the blood fuel supply following meals and between meals. Insulin and IGF-1 remove nutrients from the blood stream following meals. Insulin directs nutrients into the storage pathways that occur in the liver and fat cells. Conversely, IGF-1 directs nutrients into the vast majority of other cell types. The healthy body having at least one hundred times more IGF-1 than insulin in the blood stream evidences this fact.

Increased insulin becomes necessary to shore up lagging growth hormone with its consequent diminished IGF-1 output. There are three subtle, but dangerous consequences to a body that relies on increased insulin production. First, the stimulation of the appetite center leads to an increased tendency to gain weight. Second, increased insulin stimulates the liver in its manufacture of LDL cholesterol. Third, there is an increased reliance on cortisol, glucagon and adrenaline to keep the blood sugar elevated, between meals, when growth hormone levels fall. The consequence of normalizing blood sugar levels between meals with elevated cortisol, glucagon and epinephrine is a loss in protein conservation (muscle and organ mass). Only growth hormone helps retain body protein when fasting. Less body protein leads to less muscle mass and organ size. These are some of the major characteristics of the aging process.

Growth hormone release occurs from regular exercise, low normal blood sugars, glucagon, the low secretion rate of serotonin in the hypothalamus, and when the hypothalamus neurons secrete dopamine. There are other details, but if one keeps these five determinants in mind it encourages making better choices.

The above discussion gives a mechanistic explanation for why couch potatoes tend to develop insulin resistance. Increased insulin resistance will eventually exhaust the genetically determined ability of the pancreas to increase its insulin production. When this happens, it exhausts the pancreas beyond its genetically determined capability and adult onset diabetes manifests.

Making things worse is the fact that when GH secretion rates fall the protein content in the body decreases proportionally. GH is a fundamental requirement for the conservation of body protein between meals. Without adequate GH between meals the body increases the secretion rate of cortisol, glucagon, and epinephrine in order to maintain blood sugar levels. All three of these will activate the liver machinery that converts protein stores into sugar (gluconeogenesis).

The other extreme of health contains highly trained athletes who secrete high levels of IGF-1 secondary to increased growth hormone secretion. Increased growth hormone secretion occurs because exercise increases the fuel delivery requirements. GH is one of the main hormones that raise the liver secretion rate of fuel into the blood stream. More importantly, growth hormone spares body protein content from breakdown. In contrast, the other three-blood fuel increasing hormones, cortisol, epinephrine, and glucagon make protein fair game. Another benefit of high IGF-1 levels involves the fact that insulin requirements diminish. The highly trained athlete needs very little insulin for efficient fuel delivery into his/her exercising muscle cells because of his/her high IGF-1 levels. For this reason, exercise lowers LDL cholesterol levels. The decreased need for insulin results in a lessened stimulus to manufacture LDL cholesterol in the liver.

It has long been known that growth hormone levels decline with age. A sedentary life style accelerates this decline. Conversely, regular exercise increases IGF-1 levels secondary to increased growth hormone release.

These facts unite several health consequences of insulin resistance into a common thread of causality. There occurs a sequential decline of growth hormone and IGF-1 levels with certain diets and lifestyles, with aging. The decline of these two hormones explains some of the insulin resistance occurring as age advances and sedentary lifestyles continue. It also explains how regular exercise remedies insulin resistance by raising growth hormone and IGF-1 levels. Applying this association could save owners the unnecessary complications of diabetes and the acceleration of aging processes. Lastly, the fall in growth hormone levels brought about by a sedentary lifestyle explains why muscle and organ mass decrease with age.

Growth hormone conserves protein content. Unless owners have processes operating in their lives that encourage GH secretion, they will lose protein. Lost body protein manifests as shrunken organs, muscles, joints and skin. All the telltale signs for the shrinkage of old age processes occurring. Insulin increases when GH release declines and leads to decreases in IGF-1 levels. This creates more body fat. The typical middle-aged effect involves shrunken organs, joints, and muscles that hide their assault on body form amongst the increased fat.

An Expanded View of the Pituitary Secretion, Prolactin and its Contribution to Obesity

Within the master hormone gland (the pituitary) there resides an additional powerful hormone, prolactin. The ovary and testicles become inhibited when prolactin releases beyond low levels. Even though many physicians have been groomed into the 'knee jerk' summary that prolactin stimulates milk production, the evidence clearly implicates prolactin as a powerful inhibitor of ovarian and testicle function. The confusion arises because they know that pregnant women's steroid production becomes very high and that her prolactin level also elevates.

This false discrepancy is solved when one includes one additional fact: later in the pregnancy the ovaries are powerfully inhibited by prolactin. It is the placenta, which cranks out the huge amounts of steroids that increase with the pregnant state, despite the inhibited ovaries. The presence of a placenta during pregnancy counteracts the quiescent ovaries. However, in other pathologic states where a placenta is absent and prolactin elevates steroid production plummets.

There are four common clinical states where prolactin, produced within the pituitary, induces ovary inhibition without the benefit of having a placenta. The absence of a placenta will cause a decrease in the steroids produced by the ovary in these cases. Men always lack a placenta. Anytime a man's prolactin elevates the testicles will decrease their steroid production.

The four states that stimulate pituitary release of high levels of prolactin include: low thyroid function, birth control pill usage, chronic stress, and high serotonin. In these cases the production of the ovarian and testicle's component of body steroids greatly diminishes because these situations lead to increased prolactin. Steroids are so powerful if any one of them diminishes, the DNA content of the body will go either dormant or hyper. Most people have a pretty good feeling that when the DNA misbehaves a danger to over all health exists. With an appropriate prolactin level, healing transpires.

Mechanisms for increased pituitary release of prolactin:
- Birth control pill usage
- High serotonin levels
- Chronic stress
- Low thyroid gland function

(Note that men also can have the last three in the list, but explaining it through a women's body facilitates understanding.)

Increased prolactin levels result from the increased estrogen in birth control pills. Prolactin inhibits ovarian hormone manufacture and release. This hormone-induced mechanism associated with obesity is generally not in operation with pregnancy. The pregnant state has the growing placenta that generates needed steroids. The placenta manufactures androgens even though the ovary becomes relatively dormant by the fifth month of pregnancy. When female owners take birth control pills the body thinks it is pregnant and prolactin levels rise. Prolactin levels rise when estrogen levels approach pregnancy levels (as in with taking birth control pills).

Potential obesity occurs because like pregnancy, the birth control pills increase prolactin levels. Unlike pregnancy there is no placental hormone factory to correct the inhibition of the ovaries in their steroid production. Additional potential problems exist. Birth control pills do not contain androgens, only estrogen and progestins (abnormally shaped progesterone substitutes). Androgen production can fall and some adrenals fail in the challenge to increase androgen production and obesity ensues (**more on prolactin later**).

High serotonin levels can occur when one takes serotonin reuptake inhibitors (SSRI) type anti-depressants. The elevation of serotonin within the pituitary inhibits growth hormone release but encourages prolactin release. This fact may help explain why sexual dysfunction remains such a prominent side effect when these prescriptions are taken.

Chronic stress tends to raise prolactin levels. Prolactin levels elevate when increased cortisol release stimulates prolactin release as part of the stress response.

Low thyroid gland function will stimulate the hypothalamus initially to release thyroid-releasing hormone (TRH). TRH is a very powerful secretogogue for prolactin release from the pituitary.

Chapter 5

The Torture Chamber Diet:

Lots of Carbohydrates When One is Already Fat

Lack of attention to the consequences from abnormal hormone levels occurs with the complex endorsed diet. Abnormal hormones create abnormal urges in its adherent's feeding behavior. Abnormal hormones exaggerate feeding obsessions. The torture chamber effect describes the feeding obsessions that result from this official diet. As long as the American public believes in the diet, endorsed by the complex, there will be continued economic need from the complications of obesity. Examples of these complications are: high blood pressure, diabetes, and heart disease. The complex endorsed diet easily exposes itself when some basic missing facts return to the analysis.

A narrow band of truth occurs in the complex endorsed diet but it concerns only one type of physique. The physique for which the complex endorsed diet works occurs in an athlete who already exists at his/her ideal weight and/or peak performance. Peak performing athletes attain optimal weight and have

81

achieved optimal physical fitness. Through training, genetics, and/or age, they have the right balance of hormones.

Owners at ideal weight and physical fitness levels have properly proportioned hormone message content. This allows the proper appetite stimulation and exercise motivation to continue. They can handle increased carbohydrate intake that necessitates only a slight increase in insulin production because they have high IGF-1 levels. They tolerate slight increases in insulin because their lifestyles and/or genetics allow a sufficient amount of IGF-1 counter hormone. The counter hormone IGF-1, successfully counter weights the fat building message of insulin. As middle age approaches, most owners are not endowed with superior amounts of androgen (section two) and growth hormone, which are necessary for high IGF-1 production rates and release.

There has been little acknowledgment of basic scientific facts about the relationship of dietary choices and hormone consequences to feeding behavior. Some notable exceptions are found in the Drs. Atkin's, Schwarzbein and Sears diets. There have also been fewer acknowledgments about the circular trap that feeding behavior dictates the consequence of which hormones secrete. This explains the vicious and circular trap overweight owners find themselves in, despite earnest attempts to diet.

When obese owners adhere to the complex endorsed diet's tenants, powerful hormones release. These stimulate a preoccupation with the next feeding and a decreased ability to shed fat. It is a travesty to withhold acknowledgment that the complex endorsed diet creates a virtual torture chamber of emotional desires in the owner who expends effort to make positive health changes when his/her attempts to change are doomed. They are destined for defeat from their unfavorable hormones. Weight loss efforts become hopeless until they obtain guidance about changing their hormones for the better.

Dr. Atkins was one of the first physicians to recognize the powerful role that hormones play in feeding behavior. He did this by reviewing what was known about the hormone, insulin, in basic medical physiology textbooks. He studied cultures that do not have high rates of obesity and obesity related diseases. Dr. Atkins first began to apply what was known over twenty years ago about hormone levels and consequent feeding behavior. He correctly reasoned that insulin levels that are allowed to reach higher than optimum levels act as a powerful appetite stimulant. This creates an obsessive preoccupation for the next feeding event. He understood that insulin has a dramatic effect on the ability of the body to manufacture fat. He also understood how insulin prevents the body from accessing fat reserves.

Health benefits from lowering insulin levels are receiving renewed interest. Lowered insulin levels will decrease the stimulation in the appetite center of the brain and also increase the ability to use fat for energy. Obesity is on the increase in America and it can rarely be curtailed without an improvement in feeding behavior hormones.

It is important to extend the work of doctors like Atkins, Bernstein, Sears, and Schwarzbein. Their important contributions allow consideration of the other obesity related hormones (see earlier discussion). Briefly, the other hormones, in addition to insulin, that need to be normalized before weight loss occurs are cortisol, androgens, estrogen, IGF-1, epinephrine, and thyroid. Attention to the rebalancing of these hormones provides an extension of these authorities work and a magnification of the weight loss possible.

The importance of real food versus processed food also needs to be added into the plan. Real foods provide the mineral nutrition necessary for maximum avoidance of the complications from obesity related disease. The real food component of successful dieting explains why the opposite approach to dieting has some success. These diets for which Dean Ornish and Nathaniel Pritikin are most famous to do a better job about expressing the importance of real food in place of processed food. However, these diets fail their adherents because of the high insulin that results from the high carbohydrate content contained in these diets. In the end both diet camps on the extremes have some success, but each fails in maximizing success by ignoring either the hormones or the importance of real food.

Consuming real food (from the garden, off the tree, organic eggs, fresh fish, fresh chicken and meats that have not been canned or salted) helps avoid ingredients that are easily missed in the high protein diet. High protein dieters need to take care not to consume high protein sources from processed food. Processed food is any food that comes in a box or a can. The requirements of putting food into a box or a can necessitates adding unbalanced minerals like sodium, which deviate from body design.

The diet that has the least success in the long haul is the ADA diet. The ADA diet takes the worst features from both diet extremes. It advises consumption of 50-60% of total daily calories from carbohydrate. Carbohydrate consumption in this proportion of total daily calories condemns the obese owners into the vicious torture chamber cycle. Weight gain consequences occur because of the obligatory rise in insulin levels. The torture chamber also involves a degree of other hormone imbalances.

A discussion of the additional hormones involved in obesity brings up a concept that Dr. Atkins calls **metabolic resistance**. Metabolic resistance describes those women for whom the low carbohydrate diet proves slow to effect weight loss. This author feels that the high protein diet approach fails to acknowledge that many of these women need the added benefit of androgen. Androgen deficiency explains some of the cause for this phenomenon quite well, as it is androgens that oppose fat gain. Fat gain opposition occurs because androgens provide message content to the liver to increase IGF-1 production. Fat gain accentuates in some female owners because they have less androgen compared to men. The removal of the ovaries and the onset of the menopause can exacerbate androgen deficiency. Consideration of the 24-hour urine test for steroid production will identify this type of metabolic resistance caused problem.

Once androgens are attended to, in as safe a way as possible, these female owners need to understand how to encourage increased growth hormone release. Exercise and fasting will stimulate the capable pituitary's release of growth hormone. The few women that are unable to increase their growth hormone output by these methods may need growth hormone shots (see discussion at end of book).

An additional cause of **metabolic resistance** concerns the increased production rate of the stress hormone, cortisol. High cortisol levels in the urine identify other owners who have trouble with weight loss despite strict adherence to a low carbohydrate diets. High stress will increase cortisol release. Increased cortisol in a setting of mental stress will elevate blood sugar inappropriately and it will only come down with exercise or increased insulin secretion. It is the increased insulin secretion brought on by stress occurring in a sedentary lifestyle that leads to obesity. The intensity of the problem magnifies because weight gain prone owner's adrenal glands are particularly adept at cortisol production when they feel stress. These obese owners make more cortisol with the same amount of life stress compared to a non-obese owner. Increased cortisol causes an inappropriate increase in blood fuel that has nowhere to go in a sedentary owner until insulin releases. Only exercise and stress management will provide a way to stop this hormone cycle in the torture chamber.

Thyroid hormone levels need to be carefully evaluated. The thyroid gland determines the rate at which calories can be burned in the cell power plants (see thyroid subsection).

Estrogen levels, when high, as occurs in pregnancy and birth control pill usage, exacerbates the obesity problem in some female owners. Environmental estrogen problems can occur in men and women (see hormone mimics). High estrogen levels stimulate growth hormone release, but inhibit IGF-1 release. This aberration leads to insulin resistance because less IGF-1 becomes available to help insulin with fuel uptake by the cells.

Consequently, increased insulin becomes necessary to bring the blood fuel level back to normal. The higher the insulin levels, the more liver fat making machinery stimulated to make fat and cholesterol. More fat then becomes available for storage sites in the liver, arteries, and fat cells.

Once again the placenta rescues most pregnant women's high estrogen state by increasing the secretion of human placental lactogen, which increases IGF-1 production to 2-3 times normal blood levels. However, the high estrogen states of women on estrogen prescriptions do not have a placenta to correct the fall of IGF-1. High estrogen levels cause a fall of IGF-1 and these owners' liver make more fats from the high insulin needed to correct the deficit. Note that it is the high cortisol levels of the pregnant state that are thought to cause insulin resistance (see cortisol discussion as to why this is so).

Epinephrine release is an extension of the stress response. Like cortisol it contains message content that instructs the liver to elevate fuel in the blood stream. When mental stress occurs there becomes little need for the extra fuel in the blood stream. Eventually, insulin needs to be released to normalize blood sugar caused by mental stress's inappropriate elevation of the blood fuel. Increased insulin leads to an increased fat making message in liver and fat cells.

A 'tug of war' exists between the hormones that help shed fat and the hormones that make fat. It is worthwhile to assess whether these interrelating hormones occur in excess, deficiency, or are present in the optimal amounts needed for a healthy body.

Thin people who eat as much as they want are not always fine specimens of raging androgen production. Deficiency in muscle mass usually provides a clue that increased androgen may not be the reason for perpetual thinness. Emaciated skinniness in the presence of increased caloric intake can be due to poor digestive absorption of critical nutrients. Some owners become unable to manufacture extra insulin in a setting of increased carbohydrate intake. They never develop diagnosable diabetes because their pancreas limps along with just enough insulin and liver IGF-1 output to keep it from spilling over into the urine (see liver chapter).

All of the Popular Diets Today Miss Vital Consideration for Weight Loss (Each has Part of the Puzzle, But Not the Whole Picture)

Two different extremes in diet philosophy have been introduced. Each has a part of the puzzle that will help shed fat. Each also contains an impediment to weight loss. The best science in each diet's approach needs incorporation, while avoiding its downside.

High protein and fat with low carbohydrate diet plans are incomplete in their effectiveness because they do not contain the right mineral ratios. Their incompleteness sheds light on how some of the other diets have a weight loss effect. Some other diets have a weight loss effect because they inadvertently partially address mineral balance. The omission of mineral balance creates the hormone imbalances (higher insulin) that the high protein diet philosophy attempts to avoid. Mineral imbalance will occur any time processed food dominates in the diet.

On the up side, the high protein diet leads to lower insulin levels. In contrast, the upside of the fresh and raw food dieters is that they contain properly proportioned minerals that will better help with hormone balance. On the downside these diets contain higher carbohydrate so the insulin need becomes higher. **Higher insulin levels prove counter-productive to any diet effort.** Mineral intake balance is important to decrease the obesity hormones initially.

The Body's Mineral Design Conservation Features are Obsolete in Face of the Processed Food Diet

The human body was designed for a natural mineral ratios intake. This would be a minimum of three times as much potassium as sodium and there should be sufficient magnesium to counter calcium. **The sodium and potassium intake ratio more than reverses when one adheres to a processed food diet.** Magnesium intake is commonly deficient as well.

In prehistoric times a survival advantage occurred for anyone who could retain sodium. Natural food is relatively deficient in sodium content compared to potassium. When one eats natural food the potassium to sodium mineral ratio is greater than three to one. **Processed food has a drastically altered mineral content. Processed food diets have greatly diminished potassium and magnesium content. At the same time a processed food diet has a greatly increased amounts of sodium added to preserve the shelf life of the product. This combination causes a chronic imbalance between potassium and sodium (see mineral table).**

The same owner types that once, in prehistoric times, had a survival advantage now have a disadvantage if a processed diet is chronically consumed. These owners retain sodium inappropriately and have reversed mineral content included in the processed food diet (see mineral table). Around middle age owners on a processed food diet will develop a whole range of health consequences. This leads to six obesity related health consequences. **These owners who predispose to health consequences from the high sodium and low potassium diet were the genetically superior human design machines of prehistoric times.** In modern times, as long as they adhere to a processed food diet, they remain on a rapid self-destruct program.

Six Different Ways Obesity Propagates Secondary to the Mineral Imbalance occurring around Middle Age

1. Insulin resistance
2. Increased fat and cholesterol synthesis in the liver
3. Loss of protein content

4. Decreased steroid biosynthesis to keep blood pressure normal
5. Slower metabolic rate
6. Stress exacerbates the mineral imbalance and weight gain

All six of these factors need to be circumvented if one desires an effective weight loss rate. If one optimizes all seven hormones (section one) that lead to obesity and corrects their mineral imbalance, their diet plan becomes more complete because they now apply the best from the different diets available. Concurrently they also omit obsolete components of these other diets in light of new scientific understanding. These are important if one really wants to know what makes them fat. When owners know what makes them fat they can progress. After all, becoming aware is part of healing.

Chronic Mineral Imbalanced Diets Cause One Type of Insulin Resistance

The chronic consumption of a mineral imbalanced diet will lead to the need for increased insulin secretion. Increased insulin becomes necessary because insulin needs sufficient potassium to deliver sugar into the cells. One potassium ion helps carry one sugar molecule out of the blood stream and into a cell.

The chronic ingestion of reversed ratios between potassium and sodium leads to a decreased availability of potassium for insulin directed sugar removal out of the blood stream. The delay of the blood lowering effect of insulin leads the pancreas to secrete more insulin eventually.

The delay of potassium availability occurs after many years of consuming reversed mineral ratios. The blood stream amount of potassium contains only 2% of body potassium. The other 98% of potassium resides in the cells. It is the potassium in the cells that donates itself to keep the smaller potassium pool constant in the blood stream. Owners that eat processed foods will inevitably deplete their total potassium. The potassium in the blood stream can be thought of as the 2% 'tank' of potassium content. It is the very last tank to deplete. The standard test at the doctor's office measures the serum component in the blood stream only. The blood stream value will only change when the larger tank severely depletes. The larger tank, containing 98% of potassium resides inside the numerous cells. Cells, including red blood cells, will sacrifice their potassium content in order to keep the blood levels in the normal range.

Herein lies the deception occurring in America today: Physicians wrongly reassure their patients about the potassium levels in their blood stream while they have not inquired about the status of the larger tank. Failure to consider the consequences that ensue when the body becomes chronically deprived of the correct ratios between potassium and sodium intake leads to chronic degenerative diseases. Insulin resistance related disease is only one of several consequences of diminished potassium compared to sodium content.

Examples of chronic degenerative diseases that have a component of causality as a consequence of imbalanced potassium to sodium intake are: adult onset diabetes, high blood pressure, high cholesterol, obesity, fatigue, and anxiety syndromes.

Insulin resistance can eventually progress into adult onset diabetes (liver chapter). The accompanying signs of obesity and an abnormal cholesterol profile often associate with adult onset diabetes. The mineral balance between potassium and sodium dramatically affects all three of these processes. Misery propagates when this important relationship remains ignored.

One type of insulin resistance is caused by a chronic imbalance between potassium and sodium intake. As the mineral imbalance increases more insulin secretes to normalize sugar intake because most cells require potassium to bring sugar aboard. For each sugar transported, into most cell types, requires one potassium ion. The trouble arises from the fact that the cells donate their potassium to the blood stream when potassium occurs in scarce supply. The needed potassium donation occurs at the expense and sacrifice of the potassium content of other cells.

The pancreas senses that the blood sugar remains elevated and more insulin eventually releases when blood sugar-lowering delays. Insulin resistance describes the increased amount of insulin needed to do the same job for a specific sugar load. When an owner's pancreas exhausts in its ability to produce additional insulin, then adult onset diabetes results and blood sugar begins to rise. Many people's pancreases prove able to keep making more and more insulin and therefore they do not get diabetes. However, the high insulin levels make both of these types of owners obese and to have abnormal cholesterol levels. The only difference between the two types of owners concerns that in one the pancreas reaches exhaustion and blood sugar rises.

In both types the high insulin levels promote the blood vessels getting clogged with fat. The ability of insulin to lower the blood sugar is potassium dependent. Less potassium availability will delay the ability to lower the blood sugar level. The pancreas senses this delay and more insulin eventually secretes as the body cells sacrifice their potassium content, which insulin needs to work. The end result is that the peripheral cells receive less nutrition. They need adequate potassium to bring sugar aboard. In contrast, the liver uptakes a higher amount of sugar and processes it into more fat and cholesterol than is healthy because potassium is not needed by the liver when it converts sugar into fat. However, in order for the liver to store sugar as glycogen there needs to be a fixed amount of potassium available. Again the potassium deficiency changes the balance from the way that healthy livers store energy.

Blood vessels get fat and people get fat when the fat maker message is present. Insulin always delivers the fat maker message. If there is no insulin, there is no fat. With high insulin there will be more body fat. The potassium deficiency of middle age is one cause of insulin resistance.

Increased Cholesterol and Fat Synthesis in the Liver

An increased message content in the liver to make more cholesterol and fat occurs when insulin resistance develops. Cholesterol and fat are made from the sugar that does not enter other cells because of diminished potassium content. Diminished potassium content impedes the ability of the peripheral cells, like muscle cells, to uptake carbohydrate nutrition. The increased blood sugar becomes more liver accessible. The liver does not need potassium to suck sugar out of the blood stream and make fat and cholesterol. All the liver needs is adequate message content from insulin and it begins sucking out the blood sugar. The liver needs adequate potassium, like other cells to store sugar as glycogen. Glycogen storage requires fixed amounts of potassium to sugar. Without adequate potassium the liver is only able to make cholesterol and fat. Next, the increase in availability of sugar in the liver and the increased insulin message in the liver accelerate the liver manufacture rate of cholesterol and fat particles, LDL cholesterol. Increased levels of LDL (triglycerides) cholesterol provide a hallmark of high insulin states. This mechanism explains the potassium deficient diet's contribution to this problem.

In the insulin resistant state, at the level of the liver, the low carbohydrate diet can fail to protect the owner because no one has counseled them about their potassium deficiency. Potassium deficiency will lead to increased insulin production (insulin resistance) even on a low carbohydrate diet. If these owners restore total potassium content (it often takes 3 to 6 months), their insulin needs will drop dramatically over time.

Many owners on low carbohydrate diets do not correct their potassium deficiency and this alone causes an increase in their insulin levels. The increased insulin levels direct the liver to produce greater amounts of LDL cholesterol from the little carbohydrate consumed. When LDL cholesterol levels increase there becomes an increased risk for blood vessel disease and obesity. The increased insulin level directs both of these disease processes.

When an owner increases their potassium content they will be able to tolerate more carbohydrates without abnormal increases in their insulin levels. Obese owners are warned to initially curtail carbohydrates dramatically to decrease the appetite center activation that insulin directs. The more normal the weight becomes the more carbohydrates from real food that can be consumed.

The reason adequate potassium content in the body figures so importantly for controlling cholesterol concerns its ability to help normalize blood sugar with less insulin. Less insulin means less fat and cholesterol synthesis in the liver. The liver is a faithful servant that does as the message directs.

The tug of war between glucagon and insulin in the liver was previously discussed. The low carbohydrate diet, in the presence of adequate body potassium, will have more glucagon message content. More glucagon message content will, in physically active owners, curtail fat and cholesterol synthesis. This explains why owners on high protein and fat diets with low carbohydrate intake have decreased cholesterol levels.

The fact that one needs adequate potassium for their body to hold onto protein has been known by science for over fifty years. Decreased protein content results in shrinking muscles, organs, skin, and bones.

Mineral Imbalanced Diets Lead to the Loss of Body Protein

Owners that arrive at middle age with a history of consuming processed food diets will experience chronic protein depletion in their tissues. These bodies sacrifice cellular proteins in order to obtain sufficient potassium for the blood stream. This process takes many years to manifest. Even though their protein depletes at a slow rate, eventually these middle-aged victims begin to look typical. Usually it manifests as an increased middle area from fat accumulation and smaller muscles in the limbs and chest areas. The protein depletion also shrinks the size of their organs.

The protein depletion process occurs because potassium in the cells stabilizes the protein content. When a cell loses potassium the protein content will decrease. Little muscles, little organs, and shriveled skin result from mineral imbalanced diets because processed food contains an altered mineral content.

Less body protein translates to decreased cell function in the affected cells and less need for cell energy. Less energy equates to fewer calories needed before weight gain occurs. Less energy also means less ability to participate in what life has to offer. The cycle of obesity breaks when a middle-aged owner begins to understand how to regain mineral balance. The first step of this process involves a commitment to real food diets that restore mineral intake in their proper proportions (see mineral table).

Mineral Deficient Diets Reduce Steroid Production to Have Normal Blood Pressure

The body, which is chronically fed altered mineral ratios faces a difficult choice in middle age. It can try to maintain steroid production, but the side effect is an increase in blood pressure. Alternately, some bodies decrease steroid manufacture, but blood pressure normalizes.

Owners that eat a real food diet can secrete optimal amounts of aldosterone without raising blood pressure because they consume the right ratios of minerals. Owners that eat processed foods consume mineral ratios that destroy body functions. These altered minerals eventually strain the ability to keep an appropriate mineral balance.

The altered mineral balance causes the middle age problem in both cases. Some bodies increase blood pressure to continue manufacturing adequate steroids that depends on adequate aldosterone production in the adrenal (see below). Other bodies diminish aldosterone production, but have a normal blood pressure. Although these owner's bodies have normal blood pressure they will age more quickly because of their lowered steroid production rates. Lowered steroid production rates mean that there will be less rejuvenation message content. Less rejuvenation message content leads to the accumulation of wear and tear changes in these bodies. The more wear and tear changes accumulate the older that body looks and feels.

Aldosterone gives the message to the adrenals and gonads to increase steroid production. Owners do not tolerate increased aldosterone levels with mineral imbalances between potassium and sodium. Potassium and sodium imbalanced owners will conserve excess fluid when aldosterone becomes elevated. Excess fluid leads to blood pressure elevation.

Mineral Imbalances Lead to a Slower Metabolism

The diminished protein content slows the metabolic rate. Protein content comprises the active metabolic fraction (the part that burns calories) of body tissue. Proteins, like enzymes and mineral pumps, consume energy and therefore metabolize calories. Metabolism also slows with mineral imbalance because there is less electrical potential across mineral depleted cell membranes. Mineral imbalance slows metabolism because it diminishes cellular charge. When the body rests the majority of energy expends in recharging the trillions of cell membranes (the cellular force fields). These can only recharge adequately when the right mineral ratios oppose one another.

Each cell uses the cell membrane charge energy to sustain life. Less membrane energy content occurs when the minerals alter from their optimal proportion. This is similar to what would happen to a car battery that had its mineral content altered. As the car battery membranes mineral concentration alters so does its ability to perform useful work. Car batteries function better when the manufacture adds in the proper mineral proportions between the membranes. This fact is a prerequisite before maximal charge can occur. So it is with body cells, before they can charge the right mineral proportions need to be available.

The food industry is not cognizant of this basic body design feature. Magnesium and potassium deplete in food from processing it. Next large amounts of sodium combine into processed food in order to retard spoilage. This formula of altered mineral ratios dumps into cells year after year (see mineral table). Around middle age feeble cell batteries lead to weakness, fatigue, diminished calories burned and weight gain (*the low voltage cell syndrome*).

Stress Exacerbates Mineral Imbalance

Stress will increase the need for insulin. Increased insulin leads to an increased fat making message content. There is an additional way that chronic stress makes fat. It concerns the extra potassium loss that cortisol causes. Cortisol increases during the stress response. Increased cortisol causes increased sodium retention and increased potassium loss. This aldosterone like effect of cortisol occurs because at high levels cortisol will create message content similar to aldosterone in its sodium retentive effects. Sodium retention and potassium wasting are not a problem with normal levels of cortisol. Cortisol at normal levels is weak in its message content to retain sodium and excrete potassium.

Surgeons are well aware of this fact in post-surgery states. The body cannot survive the stress of surgery unless there is a massive output of cortisol from the adrenal glands. The increased cortisol excretion rate depletes potassium. Surgeons routinely give intravenous potassium postoperatively because the owner's body will secrete increased cortisol in order to survive the stress of surgery. The increased potassium in the IV prevents a precipitous fall in body potassium. With mineral balance in mind the real food diet that allows this is contrasted with the processed food diet.

Two Diets on Opposite Extremes in Mineral Content

Real food is high in potassium and magnesium, unprocessed and low in sodium (see mineral table for specifics).

Fresh vegetables	Fresh fruit
Eggs	Fresh meat, chicken and fish
Low salt cheese	Brown sugar
Unprocessed rice	Unprocessed grains
Unprocessed nuts	Unprocessed beans (dry or fresh)
Potatoes	Avocados

Processed food is high in sodium content, but low in both magnesium and potassium

- Anything that comes in a box
- Anything that comes in a can
- Anything from a fast food restaurant
- Anything that has more sodium content than potassium content
- Some frozen foods have sodium added
- Store bought bread, with a few exceptions
- Condiments (catsup, soy sauce, salad dressing, steak sauce, and pickles)

A general goal is to obtain over four thousand milligrams of potassium, less than one thousand milligrams

of sodium, over three hundred milligrams of magnesium and about five hundred milligrams of calcium a day.

A word of caution becomes necessary for those owners that are already overweight. Overweight owners need further dietary restriction within the real foods that are high in carbohydrate content. Even though some foods are real foods, when an owner is already overweight, the high carbohydrate foods need to be further restricted. Carbohydrate curtailment allows insulin needs to drop.

A lowered insulin need is the primary move for exiting the torture chamber. Once one moves outside the torture chamber, they can begin to lose weight. Weight loss accelerates when both the carbohydrate intake and the mineral imbalanced components are corrected.

As a normal weight approaches, there will be increased tolerance for more carbohydrate. Every owner's physiology differs and needs the counsel of a competent physician for sustained weight loss to occur.

A good place to start involves the almost complete elimination of carbohydrate contained real food and all processed foods. The high carbohydrate containing real foods are potatoes, rice, beans, grains, brown sugar, honey, and pasta. Following this initial approach will counter the IGF-1 deficit occurring in obesity by creating less insulin-produced side effects. Once the target weight, mineral balance, and hormone balance are achieved, some carbohydrates, from real food sources can be allowed. Again, each owner's physiology is unique. The counsel of a competent physician is necessary.

As a general guide keep carbohydrates consumption below 100 grams a day. Very few middle aged and Syndrome X types can consume more than this much carbohydrate a day without gaining weight and seeing their blood fats worsen.

Summary of the Weight Loss Considerations:

- Optimal hormone levels for insulin, cortisol, androgens, estrogen, thyroid, epinephrine, and IGF-1
- An exercise program to counteract cortisol and increase growth hormone

- Stress management
- Real foods diet that provides a balanced mineral intake
- Correct nutritional deficiencies (see nutritional deficiency chapter)

Case History (Application of Weight Loss Principles)

Philip was a forty-three year old professional that began to notice weight gain over the last several years. His weight gain occurred despite a vigorous work out schedule that was often one hour long for each session. Workouts included runs in the mountains; strenuous uphill climbs, and prolonged mountain bike rides. Despite his commitment to fitness training, he continued to notice a slow, but progressive 'fat tire' around his midsection. He attempted to follow the ADA diet and was always hungry. Food was constantly on his mind. He was in the torture chamber. Hormones drove his excessive feeding behavior.

Eventually Philip came across some high protein diet type books and figured it would not hurt to give this contrary advice a try. In these books it was explained how to get insulin levels down and how this would greatly diminish the preoccupation with the next feeding event.

Through my counsel, Philip eventually went on to learn about several other hormones that affect feeding behavior and the tendency to gain weight. He also began to understand the hefty contribution of insulin to his high cholesterol level. To Philip's credit he exercised regularly, which increased his testosterone production and secretion from the gonads. Testosterone counteracts the desire of insulin to make fat by raising IGF-1 production. He came to understand that even with regular exercise, middle age leads to a tendency for decreased testosterone production.

Philip began to understand how increased carbohydrates consumption increases his insulin secretion that eventually tips the scale, in the setting of falling testosterone, for increased fat manufacture. The middle-aged body tolerates less carbohydrate because IGF-1 levels have fallen consequent to a fall in testosterone levels (note: DHEA is also important here).

Marked genetic variability occurs for how much insulin one needs to stimulate the liver into excessive production of LDL cholesterol. When LDL and/or triglyceride levels elevate, as a general rule, then suspect high insulin as the culprit. Remember that both diminished thyroid function, specific nutritional deficiencies and very rare genetic defects can cause the same abnormalities of increased blood fat of this type. Strict adherence to a low carbohydrate diet will dramatically lower LDL cholesterol in most people. The status of other hormones (thyroid, androgen, cortisol, IGF-1, estrogens, and adrenaline) importantly affects any weight loss effort. Finally, mineral balance and its influence on insulin levels need to be optimized if weight and cholesterol normalization is to be realized.

Individuals with high testosterone (athletes, young adult males, and body builders) can tolerate a higher carbohydrate intake. Likewise, an owner with mineral balance between sodium and potassium can handle more carbohydrate intake. In both cases, these owners need less insulin to move sugar into their cells. Less insulin correlates with a diminished fat manufacture rate.

It is the high testosterone and growth hormone, with consequent IGF-1 increases, combination occurring in youth that allows a decreased insulin requirement. The increased growth hormone levels increase IGF-1 release, which facilitates sugar uptake out of the blood stream without the fat manufacture message content contained in the insulin hormone.

Once fat begins to accumulate the body hormones must change in order to return to a trim physique. This summarizes what happened to Philip before he realized this fundamental fact in the attainment of a more youthful physique again.

When an owner achieves his/her optimal weight, some increased (tailored to activity level) carbohydrate intake becomes allowed. Owners like Philip need to understand that by decreasing their insulin and cutting back on carbohydrates, they will decrease the stimulation of their appetite center in the brain.

As owners head into middle age they destine themselves to failure if they adhere to the ADA diet. Failure usually manifests as a slow, but steady increase in abdominal obesity measured from one year to the next. **The torture chamber always wins until a hormonal harmony facilitates weight loss again.**

Knowledge provides power to take action in the destiny of one's physique. This happened in Philip's case as he applied basic hormone knowledge his 'middle aged physique' began to rejuvenate to a closer version of his youth. He also noted a dramatic decrease of total cholesterol and LDL cholesterol (triglycerides). This means that Philip will have to watch carbohydrates more closely than others because a return to unfavorable cholesterol will always result if insulin levels again increase.

This dramatic improvement in Philip's LDL cholesterol and triglycerides occurred despite his eating four eggs with extra cheese every morning for breakfast. This effect explains the ability of the high protein and fat diet to raise glucagon while lowering insulin. The change in the hormone ratio will turn down the rate of liver synthesized cholesterol.

Philip had an added weight loss advantage by regularly engaging in aerobic exercise that burns calories but also stimulates the gonads to manufacture and release increased androgens. The ratio between glucagon and insulin will improve with regular exercise. Glucagon turns off cholesterol synthesis in the liver and increases this fuel in the blood stream. The increased fuel release that the glucagon message directs, without adequate exercise will eventually require more insulin. Viewed in this way, it becomes clearer why regular exercise proves as one of the biochemical advantages for prolonged health.

Later in the workup process, I noted high stress operating in his life. Prolonged stress depletes adrenal glands and affects the adrenal itself. It also increases cortisol release that directs energy away from cellular rejuvenation and into survival pathways. When stress is prolonged it shifts away from the optimal ratio between DHEA (an adrenal androgen with testosterone-like activity) and cortisol production. More cortisol and less DHEA production result from the chronic stress response. Increased cortisol is one of the hormones that direct the gonads to manufacture and release (by a circuitous pathway) less androgen.

The increased amount of cortisol also directs the liver to dump sugar into the blood stream. Modern stress is usually mental in nature. This extra sugar cannot be used without physical activity. When this sedentary stress occurs, insulin releases to bring the blood sugar back down to normal.

When stress becomes the operational emotion there needs to be consideration about the message content to raise blood sugar even when no carbohydrates are eaten. Now this discussion flips things on their head. As cortisol increases, the reoccurring body theme that the hormones need balance comes into play. The only difference here concerns the fact that stress creates a situation where cortisol is the weight that needs the counter weight of insulin to put the breaks on the increase in blood sugar. This fact provides another mechanism for creating a torture chamber within if prolonged stress occurs.

The fact that cortisol raises blood sugar makes sense teleologically when one remembers the advantage of increased blood fuel in physical survival situations. Prehistorically, when a human ran from the jaws of some large animal, a rapid rise in blood sugar conferred a survival advantage by increasing the physical strength. Ample fuel in the blood stream facilitates muscle fuel delivery. The problem today occurs because many stresses are psychological (mental). The predominance of psychological stress means that no flight ever comes. The stress molecules circulate and direct valuable energy inappropriately. **One of the inappropriate consequences of mental stress concerns the increased blood sugar that it causes.**

The final point about Philip was the hardest part for him to realize. Mineral imbalance in middle age plays a substantial role in fat production. Little recognition occurs between the connection for mineral balance and fat reduction. This information remains missing from many diets offered today (see above mineral balance discussion).

Philip eventually began to appreciate the many similarities between car batteries and his cells. He began to realize that he would not alter the intended mineral composition of a car battery any more than he would his trillions of cells. This relationship helped him to understand that unless he took in mineral ratios similar to body design, his many cell batteries would deplete. When minerals are consumed in the proper design ratios the cell batteries can charge. Only a body that has fully charged cell batteries can sufficiently liberate enough potassium into the blood stream to help sugar enter cells. Adequate potassium lowers insulin requirements dramatically. The mineral determinant is the fifth determinant of how much insulin a body needs to normalize its blood sugar. All five determinants were eventually improved in Philip's life.

In summary, the five basic determinants of insulin requirements are:

1. Carbohydrate load
2. Mental stress load
3. Exercise level
4. IGF-1 levels (dependent on both growth hormone and androgen levels). Excessive estrogen can also decrease IGF-1 despite growth hormone and total androgen levels being normal or high.
5. Mineral ratios of intake within the diet.

When these five basic determinants become optimal an owner will have a normal insulin level. Other factors exist (see earlier discussion on hormones), but these make up the central players of the fat manufacture rate potential in a body. They need to be reconciled first and the other factors can be worked on later.

Chapter 6

Syndrome X:

A Deadly Form of Obesity

Gerald Reaven, MD, of Stanford University, coined the term, *Syndrome X*. This term describes those owners who have embarked on an accelerated path to old age. These owner types age because their blood vessels 'rust' and then go on to develop consequent blockages. Dr. Reaven believes that Syndrome X results from having high blood insulin. He explains the clinical signs of this syndrome as the result of the high insulin state. His opinion of the clinical signs of the high insulin state are: increased abdominal fat, high blood pressure, increased blood triglyceride level, elevated LDL cholesterol, increased skin tag growths on the neck and under the arms, increased blood clotting tendency and an accelerated rusting [oxidation] rate within the blood vessels.

The author of this book feels that the signs of high blood pressure, increased blood clotting tendency and increased rusting rate are better explained by three additional factors. The first concerns the chronically elevated production of cortisol caused by a stress filled lifestyle. The second involves the fact that these owners tend to consume a processed food diet. The third factor and perhaps more central to the underlying cause of this syndrome concern the fact that they have a diminished IGF-1 level.

Recognizing Syndrome X:

1. Look in the mirror and note whether or not their trunk is their predominant area of obesity. High insulin levels deposit fat here preferentially.

2. Develop more suspicion, if their cholesterol and triglycerides are elevated. High insulin secretion rates drive the elevation of cholesterol and triglycerides in most owners.

3. Obtain a fasting insulin, C-peptide, and IGF-1 level. Many physicians are misled when the fasting insulin comes back normal. They fail to realize that insulin's half-life decreases with a diminished IGF-1. An elevated C-peptide catches these cases of occult Syndrome X. In addition, measurable insulin in the fasting state is abnormal (see below as to why).

4. Skin tags are new moles, suspended by a narrow stalk, most pronounced on the neck and armpits. Insulin is felt to be a growth factor for these moles.

5. High blood pressure adds further suspicion. High insulin levels leads to increased levels of blood pressure elevating hormones and decreased levels of blood pressure lowering hormones (see below).

6. Appearing older than their chronological age because Syndrome X patients tend to motivate by stress. Stress increases the fight or flight emergency response hormones and decreases the repair hormones (see below).

7. Because of the increased stress response operating within the Syndrome X owner's life, the blood-clotting tendency elevates. Specifically, the stress response causes an elevation in the acute phase

reactants. One of the acute phase reactants is called fibrinogen. Fibrinogen, when elevated, increases the tendency for blood vessels to clot from within (explained in the physician's side bar below).

Syndrome X is also known as metabolic Syndrome and insulin resistance Syndrome. This syndrome has been attributed largely to increased insulin secretion rates. Elevated insulin levels describe only one of the hormone aberrations causing Syndrome X. This syndrome becomes more interesting when the other hormone abnormalities are included.

Exaggerated insulin release rates, following mental stress, or following a carbohydrate meal are only required when IGF-1 levels begin to fall. Healthy individuals have 100 times IGF-1 compared to insulin levels within their blood streams. Because IGF-1 has a half-life of four days, once it releases into the blood stream, it hangs around much longer than insulin's five to ten minute lifespan. When either or both androgen or growth hormone levels begin to fall the liver releases less IGF-1. IGF-1 occurring at high levels greatly reduces the amount of insulin released because it facilitates the body cells outside of the liver and fat in their procurement of nutrient uptake. As IGF-1 levels begin to fall off, secondary to sedentary lifestyles, chronic stress, frequent feedings or the poor hormonal levels mentioned above, more insulin release becomes necessary. More insulin becomes necessary because the peripheral cells outside the liver and fat have less IGF-1 to assist in their nutrient uptake.

Insulin however contains message content that IGF-1 does not share. The insulin message directs the liver to manufacture increased amounts of sugar into both fat and cholesterol. IGF-1 message content does not direct this process. The progressive inverse changes in the amounts of a Syndrome X individual's insulin versus IGF-1 levels provides a more complete explanation for their primary hormonal defect.

When IGF-1 levels remain high the body cells, outside of the liver and fat, procure an increased proportion of blood fuel. However, because of the anatomical location of the pancreas dumping insulin directly into the portal vein, the liver is positioned to always receive the highest amount of insulin message content. A falling IGF-1 level tips the advantage to the liver procuring more fuel for the manufacture of cholesterol and fat.

Fact: the higher the insulin and the lower the glucagon level within the liver the more active HMG CoA reductase. HMG CoA reductase is the enzyme, which makes cholesterol in the body. This detail explains how the high insulin states like adult onset diabetes, and Syndrome X develop abnormal cholesterol profiles.

Once again, instead of the medical industrial complex revealing this simple scientific relationship, the more lucrative cholesterol lowering drugs fill media advertising space. How one heals from Syndrome X involves an emphasis for the methods that will increase one's IGF-1. Improvement here leads to an improved cholesterol profile without side effects or toxicities.

The overall genesis of the typical Syndrome X individual results from his/her high insulin only after his/her IGF-1 levels begin to fall. Once IGF-1 falls, either the stress response or increased carbohydrate consumption leads to an exaggerated release of insulin. Increased insulin release preferentially tells the liver to make sugar into fat and cholesterol. Insulin levels become even higher in those owners who chronically subsist on a processed food diet. Processed food diets provide reversed mineral content that deviates from body design. Reversed mineral intake leads to insulin resistance around middle age. In addition, the vitamin deficiencies, which are inherent consequences of subsisting on a processed food diet, prevent adequate blood vessel repair and blood fuel removal processes (explained in nutritional deficiency chapter).

Syndrome X owners are on an accelerated track to an old body. Their fundamental defect concerns their elevated insulin levels, which become necessary because of their falling IGF-1 levels. This syndrome worsens in a setting of increased stress. Increased stress will accelerate potassium loss, as well. Low potassium and low magnesium diets with elevated sodium intake will increase the disease process (see stress and mineral depletion). In many ways a high stress hormone output could help explain how Syndrome X patients age so quickly and have high blood pressure as well. This is better illustrated by the fact that chronically elevated cortisol tells the body that an emergency situation chronically exists. When the body perceives an emergency (real or imagined) the body's energy redirects into survival pathways (catabolic instead of anabolic). In syndrome X owners the survival pathway becomes the norm instead of rejuvenation activities.

When body energy chronically directs into survival pathways, wear and tear changes will become more likely. **Wear and tear changes within the body cells manifest clinically as an old appearing and feeling body.** Wear and tear changes become more likely secondary to the lack of cellular repair activities. Cellular repair and rejuvenation activities cannot occur unless sufficient anabolic message content reaches the cells. This helps to explain why Syndrome X owners tend to age so quickly. Not only do they tend to have high insulin derived disease but also high cortisol derived disease, as well. Cortisol directs body energy into catabolic pathways. Too much catabolism within the blood vessels leads to wear and tear changes.

After all, in one way or another, the initial blood vessel lesion (rust) grabs hold when the blood vessel repair processes fail to keep up with the injury rate (excess iron or fluoride ions). The degree of anabolism determines the repair rate. Conversely the amount of catabolism puts repair on hold and survival pathways activate instead.

The severity of the Syndrome X is exacerbated by the consumption of processed foods.

Large amounts of insulin within the blood stream require large amounts of cortisol to effectively counter insulin's behavior of moving every last sugar molecule out of the blood stream. Remember insulin levels need to rise abnormally only when IGF-1 levels fall. Almost always the fall in IGF-1 levels has a component of causality in a diminished growth hormone secretion. Bodies that cannot produce sufficient growth hormone need increased insulin and increased cortisol in order to maintain their blood sugars between meals (see earlier discussions if this is not clear as to why).

Cortisol counters the effect of insulin by increasing the blood sugar levels. Cortisol, however, directs body energy and molecular building parts into survival pathways and out of repair and rejuvenation pathways. Anti-oxidant synthesis and blood vessel repair defer with a chronically high cortisol level.

Modern life stress further exacerbates the increased need for insulin. When the owner also experiences chronically high levels of cortisol release, the body energy directs away from cellular rejuvenation activities. Syndrome X causes the inefficient use of molecular building parts for repair activity by the body. The correction of this dysfunctional process begins when the healthy and appropriate ratio of cortisol and insulin returns to normal within the blood stream. This can be substantially accomplished by nutritional rather than symptom control treatment plans. Before a description of those strategies occurs a syndrome

> **Physician's Sidebar**
>
> Dr. Reaven is right about insulin having some role in the blood pressure elevation of these owner types. One of the reasons that elevated insulin raises blood pressure concerns the fact that it powerfully stimulates the blood vessel lining cells production of endothelin. When endothelin production increases blood pressure then rises. High insulin levels also diminish the ability of the blood vessel lining cells to produce nitric oxide. The diminished production of nitric oxide further reduces the body's ability to maintain an appropriate blood pressure level.

closely related to Syndrome X will be described. This syndrome can be treated by very similar nutritional strategies used for the treatment of Syndrome X.

Physician's Sidebar

Syndrome X owners have long been known to have an increased prevalence of gout. Mainstream textbooks often fail to explain why this association occurs. If one adds back in the protein sparing effect of growth hormone it begins to become clear. Additionally, concerns the fact that unhealthy owners need increased cortisol, glucagon and epinephrine to keep their blood sugars elevated between meals because their growth hormone levels have fallen off. These two facts, taken together, elucidate why gout risks increase with syndrome X.

Syndrome X owners have an accelerated rate of gluconeogenesis. Gluconeogenesis creates increased nitrogen waste that normally releases as urea and only small amounts of uric acid. However, with the high carbohydrate production rates, caused by increased gluconeogenesis, the uric acid production rate goes up, as well. Some owners have a weakened genetically determined ability to convert excessive uric acid production into urea and gout ensues.

Rather than give these patients toxic prescriptions to treat their symptoms of gout, it seems worthwhile to spend a little effort on getting their growth hormone levels back up to par (see text for methods). This approach becomes even more critical when one adds in the knowledge of the association between body protein content maintenance and youth preservation. Without someone helping these syndrome X owners to regain their protein content they are destined to age at accelerated rates.

The type A personality is a closely related condition typified by a hard driving (intense), hurried, over achieving and aggressive type of individual. These personality traits have long been associated with an increased propensity for acquiring heart disease. The trouble is that the mainstream education stops at a superficial level of explaining this association. However, a deeper hormonal explanation exists for why these types of individuals develop the inflammatory type of heart disease at an accelerated rate.

All body hormones contain message content inherent in their precise shape. All hormone message content concerns how the body becomes directed to spend its available energy. The stress response hormone's release is quite active within the type A individual. The body cannot discern whether a stress is real or imagined. In addition, once the stress response initiates body energy redirects into surviving a physical stress. This largely outdated hormonal response has health consequences in the modern world because most stress has become mental in nature. There are three big health consequences within the blood vessel that result from chronic mental stress (type A personality). These mechanisms are included in the side bar for the reader's doctor to review because the second consequence is generally considered to be above the layman's level of how to heal from obesity.

Physician's Sidebar

First, the survival message contained in the hormonal stress response directs body energy away from rejuvenation activities and into mounting energy for increasing physical strength. Ongoing blood vessel rejuvenation activities are crucial if one is to avoid the consequence of disrepair damage and eventual inflammation. Survival mode that occurs occasionally is without consequences. However, some personalities are always in survival mode. The body cells are faithful servants to the message content that they receive.

The opposite extreme of message content occurs within the blood vessels of those owners who are happy and physically active. Regular physical exercise and effective stress management promote the increased release of the rejuvenation (androgens) hormone's message content. Within the blood vessel the rejuvenation hormone's message content tells these individual's cells that it is important to invest in repair and regeneration activities. Ongoing repair activities are necessary to prevent the wear and tear changes within the blood vessels, which lead to inflammation.

The second way that the stress response leads to blood vessel inflammation, when it becomes chronic, involves the fact that certain inflammatory proteins release. C-reactive protein is only one of these types of proteins released within the stress response. Collectively these proteins are known as the acute phase reactants. Other acute phase reactants include: fibrinogen, complement, interferon, haptoglobin, ferritin, and ceruloplasmin. The increase of these protein types in a setting of physical stress makes sense. The physical stress of combat or running from a large animal requires the increased activity of the immune system, blood clotting and remanufacture of new blood cell components. Each of the above acute phase reactants contributes to this overall scheme for surviving physical trauma. However, in the setting of chronic mental stress when these acute phase reactants increase, the increased tendency for blood vessel inflammation occurs.

The above cause and effect relationship of the acute phase reactants and chronic stress develops a more comprehensive picture. An elevated C-reactive protein is only a small part of the overall picture of blood vessel inflammation. Other specific acute phase reactants like ferritin and fibrinogen contribute directly to blood vessel inflammation.

Healing Type A personality individuals' blood vessels becomes possible when they understand the larger connection of scientific facts. Rather than following the well-worn path of the hopeless mantra about the cruel hand that genetics has dealt, so "take your cholesterol lowering prescription," healing becomes possible. Healing occurs when a Type A personality sees the importance of increasing positive emotions and physical activity level while decreasing stressful behaviors.

The third facet for how the stress response causes blood vessel inflammation, which occurs chronically in the type A personality, involves the obligatory increase in blood fuel. This third facet concerns the outdated survival response, which directs the massive dumping of fuel into the blood stream. Extra fuel in the blood stream confers a survival advantage to output the increased strength required to survive physical stressors. However, mental stress also causes release of increased blood fuel. This response occurs because the body cannot discern the difference between mental and physical stress. The body is smart enough to eventually figure out that the mental stress-caused increases in blood fuel has no where to go. Increased amounts of insulin then secrete from the pancreas to direct the uptake of the unneeded fuel into the liver. This fact unites the Type A personality to the Syndrome X owner for how each of their blood vessels become inflamed.

There exists one key difference between the classical type A personality and the classical Syndrome X owner. The syndrome X owner's primary defect results from a fall off in their IGF-1, usually from either or both a fall in androgens, growth hormone secretion rates and/or liver injury. In other words, glandular failure originating in the pituitary, adrenal, liver or gonads brings about a diminished IGF-1.

In contrast, the classical Type A personality suffers a drop off in his/her IGF-1 levels because stress chronically elevates his/her blood sugar secondary to elevated cortisol. Remember that growth hormone cannot release until the blood sugar begins to fall. Here lies the fundamental pathology obstructing longevity in a Type A personality. Their IGF-1 falls off when their growth hormone fails to release because the stress response chronically elevates their blood sugars. This association makes it clearer why exercise becomes so important to restoring health to the Type A personality body. Exercise provides a mechanism to draw down the blood fuel levels and hence growth hormone release can occur.

Finally, in order to complete the introductory discussion of syndrome X, the importance of IGF-1 levels needs further mention. Healthy owners predictably have high normal IGF-1 levels. IGF-1 is synthesized within the liver in response to sufficient DHEA (simplified). Growth hormone secretion from the pituitary causes the IGF-1 stored within the liver to be released. Healthy owners have at least 100 times the IGF-1 in their circulation as they do insulin. IGF-1 acts like insulin for the cells outside the liver and fat. High IGF-1 levels lower the amount of insulin needed within the body. IGF-1 levels provide a mechanism for fuel uptake by cells outside the liver and fat cells between meals. In contrast, insulin's design facilitates the liver and fat cells to remove fuel out of the blood stream following meals.

Troubles start around middle age in sedentary and stressed owners. These two lifestyle traits combine to diminish IGF-1 levels. A lower IGF-1 level means that insulin secretion must rise to abnormally high levels in order to attempt to offset the decrease in IGF-1 levels (thus keeping the total amount of fuel nozzle hormones constant).

This detail explains why Syndrome X types present with elevated fasting insulin and/or C-peptide levels. Healthy owners have essentially no insulin in the fasting state. Most mainstream labs allow some insulin in the fasting state before it is considered too high. However, this belief proves inconsistent with what role insulin serves when one remains healthy.

The body is smart and consistent. Insulin's design helps to store fuel in the liver following meals so that there will be enough fuel released from the liver between meals to keep the blood fuel constant. However, a rise in IGF-1 facilitates the uptake of nutrition between meals that growth hormone release directs, while sparing body proteins usage for fuel. Remember that growth hormone release also directs the liver to simultaneously release sugar and fat into the blood stream that was stored in the liver following the last meal under insulin's direction.

The healthy body relies on sufficient insulin following meals to store adequate fuel for the between meals state. In the between meals state, sufficient growth hormone release tells the healthy liver to release the stored sugar, fat and IGF-1. In this way, the healthy body's cells have access to sufficient blood fuel at all times while protecting their protein content from combustion for usage as fuel.

Unhealthy people in contrast do not release sufficient growth hormone. Instead they release excessive cortisol, glucagon and epinephrine between meals in order to keep their blood fuel elevated. However, body protein now becomes fair game for fuel usage and because IGF-1 levels are down then more insulin needs to be secreted between meals. Insulin secretion in the between meals state is abnormal. It only becomes necessary when IGF-1 levels have fallen.

Physician's Sidebar

Why would a healthy body instruct the liver to draw down the blood sugar further in the between meal state? Unhealthy bodies do this only because they lack fuel nozzle hormones (a fallen IGF-1 level) for their cells, as in muscle and organs. In order to keep alive they accept the complications of increasing insulin output enough to spill over into the general circulation and allow these cells their fuel nozzle. The major complication results from the fact that whatever the insulin secretion rate, because of the pancreas-portal vein connection to the liver, the liver receives the highest concentration of its message content before the other body cells can get any 'insulin fuel nozzles'. This means a contradictory message occurs within the unhealthy body's liver during the fasting state.

The contradictory message within the unhealthy body results from the fact that insulin, within the liver, opposes the other body cells procurement of nutrition between meals (fasting state). Remember that the liver and fat cells have the highest insulin receptor concentration per cell at about 200,000. A high insulin release rate becomes necessary to get insulin beyond the liver and out into the general circulation but this also excessively stimulates the cholesterol and fat making machinery. This fact leads to the liver performing counter productive tasks during the fasting state. Some liver cells under the influence of insulin begin sucking sugar out of the blood stream, while the body is in the fasting state and thus worsen the falling blood sugar. Other liver cells under the influence of the counter hormones like cortisol and epinephrine tell other liver cells to dismantle protein (gluconeogenesis) to make sugar and release it into the blood stream. The end result accelerates protein degradation and increases the manufacture rate of fat and cholesterol. These side effects occur just to keep the unhealthy body's blood sugar elevated between meals.

Remember that normally, between meals, the healthy liver responds to growth hormone's release by simultaneously releasing IGF-1, sugar and fat into the blood stream. Growth hormone's release also protects body protein content between meals. Healthy bodies therefore avoid competing messages at the level of the liver between meals (fasting state) and conserve organ and muscle protein stores, as well.

Insulin has a half-life of five to ten minutes. IGF-1 has a half-life of about four days. For the reasons mentioned above, measurable insulin in the fasting state should raise the suspicion about Syndrome X. Fasting insulin and C-peptide therefore provides a pretty good marker for diminished IGF-1 levels (excluding diabetics). The primary health effects when IGF-1 levels fall and insulin levels rise concerns pancreas strain and the activation of the fat and cholesterol making machinery within the liver. In addition, these owners combust their body protein content for fuel needs. Not only are these owners becoming fat making machines, they also work their pancreases to death and dismantle their muscles and organs, as well.

Chapter 7

The Last Three Hormones of the Basic Seven That Centrally Influence Obesity

High Estrogen Levels Can Promote Obesity

Estrogen: a steroid hormone made in the adrenals and ovaries. Very small amounts are also made in the testes. The estrogen message has a component in instructing cells to divide. High estrogen levels also initiate a circuitous message to increase body fat when its levels become too high compared to androgens (testosterone and DHEA).

When estrogen rises beyond normal levels there occurs a varying tendency to promote two of the hormonal factors creating obesity. The ability of estrogen to raise insulin and lower androgens, in certain females, describes these two factors.

Three main clinical situations promote estrogen induced weight gain. Not all female owners will express these tendencies equally. This variability may have a genetic basis. Not all women with increased estrogen states tend to gain weight equally. High estrogen states tend to promote weight gain in many female owners. This fact will be the focus of this subsection.

The first clinical example for estrogen induced weight gain results from birth control pills. They predictably increase insulin in the body. The first mechanism for this situation arises from an abnormal hormone tandem that high estrogen levels cause.

The first part of the hormone tandem involves the high estrogen induced increased release rates of growth hormone from the pituitary gland. This occurs in the increased estrogen states that result from birth control pills usage. Growth hormone will initially raise the blood sugar level. It is the second hormone in this tandem, that estrogen simultaneously inhibits, which alters the normal pattern of events.

When estrogen levels remain optimal, the release of growth hormone directs the simultaneous release of insulin-like growth factor (IGF-1), the second hormone in the tandem, along with the liver stored sugar and fat. IGF-1 has powerful insulin-like blood sugar lowering message content. This message content lowers the blood sugar that the growth hormone released initially elevated. Here, extra insulin is not needed.

Normally, this tandem hormone effect provides an effective way for the cells to receive fuel from the blood stream, between meals, without raising insulin levels. High estrogen states, although they initially stimulate growth hormone release, counteracts the normal tandem by inhibiting insulin-like growth factor release (IGF-1). **The normal hormone tandem interrupts because high estrogen causes the simultaneous inhibition of IGF-1 release.**

IGF-1 is an insulin-like hormone that acts normally in the cells outside the liver and fat. Its presence lowers the amount of insulin needed by the body. The IGF-1 released assists insulin by taking sugar out of the blood stream and into the cells outside the liver and fat. When IGF-1 levels diminish, the growth hormone released directs increased sugar to be released into the blood stream. More insulin then has to be secreted from the liver to make up the deficit of total fuel nozzle hormones. The more insulin secreted, the more message content there is to make body fat.

The second clinical situation of estrogen-caused obesity involves increased prolactin levels caused by the increased estrogen state of birth control use and/or stress filled lifestyles. Prolactin inhibits ovarian androgen hormone formation and release. This hormone-induced mechanism, associated with obesity, generally does not operate in pregnancy because this physiologic state possesses a growing placenta. The growing placenta, in the pregnant state, more than offsets the prolactin induced inhibition of the ovaries by serving as a hormonal factory for steroid production. Birth control usage fools the body into thinking it is pregnant. Prolactin levels rise in response to either increased estrogen or cortisol. However, in either of the above cases, there is no placental hormone factory to make up the prolactin caused androgen deficiency.

In a real pregnancy, the placenta manufactures androgens even though the ovaries become relatively dormant by the fifth month of gestation. When a female owner takes birth control pills, the body thinks it is pregnant. During birth control pill usage, prolactin levels rise because estrogen levels approach pregnancy levels. However, with birth control pill usage there is no placenta to manufacture the androgen lost when the ovaries become inhibited by excessive prolactin.

Potential obesity problems occur because, like pregnancy, the birth control pills increase prolactin levels. Unlike pregnancy, there is no placenta (hormone factory) to correct the inhibition of steroid production within the ovaries. The potential for problems compound due to the fact that the birth control pill does not contain androgens, only estrogen and progestins (synthetic progesterone substitutes). Androgen production can fall within these owners and their adrenals are left all alone for this task. Some female's adrenals fail at the challenge of increased androgen production and obesity ensues.

The third clinical situation of high estrogen-induced obesity is beyond the level of this discussion. For those who are curious, it involves the dramatic increase of sex hormone-binding globulin that high estrogen levels direct. Androgen that may be produced by the ovary or adrenals, in high estrogen states, gets trapped on a carrier protein in the blood stream at 98% of the efficiency level.

Hormone Mimics

Hormone mimics: environmentally present chemical compounds that resemble body hormones enough that they deliver message content when inside body tissue. Message content from these mimics are excessive, deficient, altered and/or prolonged. These properties alter the recipient owner's normal physiology.

Hormone mimics diminish the health of adrenals and gonads. Pervasive and insidious effects of hormonal mimics surface in the worldwide phenomenon of falling sperm counts in men. Androgen levels primarily determine a man's sperm count. Hormone mimics surface from environmental and nutritional sources. Around the world chemicals like DDT and Agent Orange have been implicated in the estrogen mimic effect. These mimics compete with men's androgen tone causing sperm counts to drop progressively each decade.

The numerous common chemicals that exert a biochemical message effect along various degrees of estrogen mimicry sounds fantastical. When one considers that all estrogen mimics share a similar molecular shape with their estrogen counter part, it becomes conceptually consistent. The common shape in the 'key' of estrogen is technically called the aromatic ring. Trouble starts because these diverse chemicals all contain the estrogen 'key' shape and once inside the body they lead to an unnaturally high estrogen message content.

In many ways, estrogen and estrogen mimics counter the effects of androgens like testosterone. The most frightening aspect of turning up the estrogen message in both men and women concerns their powerful cell division message that certain estrogen types and estrogen mimics deliver. This can cause abnormal growths in the prostate tissues of some men - a prime cause of benign prostatic hypertrophy (BPH). In women, estrogen dominance (relative to progesterone) leads to an increased tendency towards developing fibrocystic breast disease, uterine fibroids, breast cancer, PMS, and others.

Saran wrap, plastic food containers, and the liners inside canned foods all contain estrogen mimics. As a general guide, the higher the fat content, the greater the tendency for hormone mimics to migrate into the food. Prime environmental hormone disrupters are DDT, PCB's, and dioxin that are widely dispersed throughout the environment.

DDT (a suspected carcinogenic in mammals) is an environmentally persistent insecticide that causes fragile and broken eggshells in wild birds. Estrogen-like substances stimulate cell growth in estrogen sensitive tissue.

Though banned in the late 1970's, PCB's used in transformers and other electrical components still persist throughout the environment. Another common source of continued exposure is from foreign grown produce and fruits, which still contain the outlawed insecticides. The list of detrimental effects includes: severe birth defects in people and animals, cancer in animals, as well as a link to intellectual deficits in children. These mimics clearly tend to stimulate cell division inappropriately.

Dioxins have been shown to cause cancer in both animals and humans and have acted like estrogen in animal studies. In 1979, the Environmental Protection Agency (EPA) banned some herbicides because they were contaminated with dioxins, but there are still numerous additional sources including paper bleaching facilities, polyvinyl chloride factories and trash incinerators. The EPA and industry modified industrial practices with some success, but it has proved difficult to eliminate them all. If consumers began a boycott of plastic and bleached paper products, where possible, the level of these disrupters would fall environmentally.

Plasticizer compounds may leech from landfills into the environment, but do not seem to linger in human bodies. Two types of plasticizers suspect in causing problems are phthalates and adipates. In lab animals phthalates cause liver cancer and testicular damage. Adipates in animal studies link to shortened life spans and decreased fertility. Bisphenol A, a building block of plastic manufacture used in dental sealants and in food can liners causes enlarged prostates in animals.

There are several low cost strategies to avoid ingesting estrogen mimics. Reduce consumption of suspect compounds by avoiding plastic containers and wrappers. Consider using alternatives to pesticides and insecticides on both lawn and pets. Wash fruits and vegetables thoroughly or buy organic foods. Limit consumption of suspect fatty foods where these compounds accumulate in the food chain. Watch for local fish pollution possibilities. When reheating food, don't use plastic. Heat accelerates transfer of the hormone mimics into the food (Consumer Reports, Aug 98).

Adrenaline Deficiency as a Cause for Obesity

Adrenaline carries out the receptor activation initiated by both cortisol and thyroid message content. Adrenaline (epinephrine) activates the receptors that these two more powerful hormones directed to be manufactured. Only when there are sufficient receptors can adrenaline's message be heard by the target cells. When adrenaline releases under these normal circumstances, metabolism increases. When metabolism increases, fuel consumption increases. In addition, sufficient adrenaline proves as a powerful inhibitor for insulin release. Because the popular weight loss nutriceutical, ephedra, is molecularly similar to adrenaline, this explains its continued demand.

Some owners suffer from various forms of adrenal insufficiency. It is of little use to manufacture sufficient thyroid hormone and cortisol when the body experiences inadequate adrenal function from the lack of adrenaline. The manufacture of thyroid hormone takes place in the thyroid gland. The manufacture of cortisol takes place in the adrenal cortex. When the cortex fails, the diminished ability to make one or more adrenal steroids becomes the problem. Adrenaline manufacture needs to follow up on what both the thyroid and cortisol message started in the creation of adrenal receptors.

Overtaxed adrenal glands can often masquerade as a thyroid problem. These owners walk like a thyroid problem and talk like a thyroid problem, but they are not a thyroid problem. Patients feel lousy and intuitively sense something is wrong. These owners' standard fatigue work-ups come back normal at their doctor's office. A superficial inquiry leads to superficial platitudes and the statement that nothing is wrong. Some of these owners end up on antidepressants. How can this be? Further intrigue occurs when one includes the many depressed and overweight owners who note a marked lifting of their depressive like symptoms while taking ephedra (probably the real reason it gets bashed as dangerous while the more dangerous drugs, like aspirin, sells over the counter).

A common example is revealed when one understands where adrenaline (epinephrine) derives from. Twenty different amino acids occur that, when arranged uniquely in sequence, type, and amount become the various proteins. The body manufactures most of these de novo (from scratch). Eight essential amino acids must be obtained in the diet. The egg provides the only food source that contains all eight. All other protein sources are deficient in one or more amino acids. Adrenaline derives from the essential amino acid phenylalanine.

The above discussion explains the importance of attaining adequate phenylalanine in the diet but there is a subtle and often overlooked reason for adrenaline deficiency.

Protein disassembly requires adequate stomach acid and digestive juices if essential amino acid supply lines are to be maintained. Owners who lack sufficient stomach acid and/or digestive juices tend to become deficient in essential amino acids necessary for the reactions of life. Deficient adrenaline manufacture describes one of the problems that can ensue due to ineffective disassembly of protein.

Another cause of adrenaline deficiency concerns the failure to obtain the necessary molecular building blocks for its manufacture. Most adrenaline is made from phenylalanine or the closely related amino acid, tyrosine. Physicians understand this much. Vitamins and cofactors, which are necessary to make either one of these amino acids into adrenaline, often go unnoticed. Deficiency of any one of these halts this critical hormone's biosynthesis. The nutrients necessary for the manufacture of adrenaline are: tetrahydrobiopterin (folate derived), vitamin C, vitamin B6, vitamin B12, folate, methionine, and SAMe (S-adenosyl methionine). The most common deficiency arises from a deficiency of SAMe which clinically evidences by an increase of blood homocysteine levels. SAMe is part of the important methyl donor system and is explained in the digestion section.

A deficiency in one or more vitamins and cofactors results in health problems. These additional health problems develop because the adrenal secretes partially manufactured adrenaline (dopamine or nor-adrenaline) into the blood stream. When this occurs, deficiencies of one or more vitamins exist. Dopamine and nor-adrenaline (norepinephrine) have different shapes, therefore, a different message. A different message results when altered molecules bind to the adrenaline-like receptors.

When elevated amounts of nor-adrenaline enter the blood stream instead of adrenaline, a drastic increased tendency toward high blood pressure occurs. A single, simple vitamin deficiency can physiologically be the cause of elevated blood pressure. Some of these hypertensive owners may prefer taking vitamins. In this case, vitamins instead of blood pressure medicine lead to healing without the predictable side effects (see *The Body Heals*).

Insulin-like Growth Factor Type 1 (IGF-1)

The Seventh Hormone to Consider for Healing Obesity

Insulin like growth factor type-1 has been discussed earlier in relationship to how it lowers total body insulin requirements. It also was earlier discussed how it keeps the fat making machinery, within the liver, quiescent. Before leaving the glandular component of obesities' propagation a few summary statements need to be made about the important role IGF-1 has towards the body cell's nutrition status.

Physician's Sidebar

When was the last time a mainstream medical doctor inquired about these possibilities before prescribing medication? Please note, it is not the intent of this discussion to cast disparaging remarks on the many caring and kind physicians in practice today. It has been quite a shock to this author, while researching this work, the numerous holes in physicians (and this author's) educational exposure addressing healing versus symptom control. Usually the physician is not the one to blame, but rather, the way that simple concepts continually and knowingly continue to be withheld. The motive is money.

Insulin like growth factor type-1 [IGF-1] occurs at levels 100 times insulin levels when an owner remains healthy. IGF-1 acts like insulin outside of the liver and fat cells. The more IGF-1 in the circulation, the more the body cells outside of the liver and fat can uptake fuel and nutrients without excessive insulin. Growth hormone release promotes the healthy liver to release IGF-1 into the circulation. Androgens, like testosterone and DHEA, tell the liver DNA programs to activate and increase IGF-1 synthesis rates. In this way when an owner remains healthy the ample IGF-1 circulates within the blood stream, which keeps his/her cells filled with nutrition.

As long as both a sufficient amount of liver manufacture and release of IGF-1 occur into an owner's blood stream daily, there will be less need for insulin release. Insulin release in healthy owners falls to very near zero between meals. However, unhealthy owners will tend to need massive increases in their insulin levels between meals because IGF-1 levels have fallen.

When IGF-1 levels fall, increased amounts of insulin are needed outside of the liver and fat so that other body cells can receive nutrition. The trouble with insulin level increases, to make up for the nutritional message deficit within the other body cells, regards its additional fat maker role. Unlike IGF-1, insulin delivers a powerful message within the liver and fat cells to uptake excessive sugar and make it into fat and cholesterol.

Healthy owners gain weight less readily because their cells, like muscle and heart, can receive nutrition with the help of IGF-1. Conversely, unhealthy owners have diminished IGF-1 levels so they will need excess insulin in order for their cells to uptake nutrition. High insulin levels make an owner fat.

Healthy insulin levels cause the liver to store just the right amount of fat and sugar within the liver and fat cells following meals. Between meals (fasting or exercising), in healthy owners, this stored fuel and IGF-1 releases under the direction of increased growth hormone. In this way the healthy owner has sufficient blood fuel and fuel nozzle hormones to feed his/her cells from one meal to the next.

The liver chapter explains more of the particulars on how important a high IGF-1 level is in regards to body health. Here it is only important to understand how when IGF-1 falls, the insulin level needs to rise. Insulin needs to rise because when one fuel nozzle hormone's level decreases the other needs to increase. Insulin on the rise means fat making will also be on the rise. Conversely, IGF-1 levels, at healthy levels, facilitate muscle and organ development and nutrition and greatly diminish the amount of insulin needed.

The Ovary and Body Fat

The healthy female's ovaries make appropriate amounts of androgens. In addition, the healthful state typifies by proper levels of estrogens, which are cycled with proper levels of progesterone. Each month a new cycle begins and her cells receive the right amount of message content, which directs them to invest in rejuvenation activities. Rejuvenation activities keep the body young.

Around menopause the ovaries begin to fail and give up their steroid production role to the middle aged adrenal. Many years prior to this, steroid production begins to diminish within the ovaries. When clinicians miss this or treat with aberrant shaped hormones, a female owner accelerates on the path to an old body. (*Lee, 1999*)

Nutritional deficiencies play a major role in ovarian disease. Common factors include a chronic preference for the consumption of a 'processed food' diet versus a 'real food' diet (see earlier discussion). Ovarian steroid production provides an example of how nutrition affects ovarian health. One central facet of ovarian function determines the rate at which cholesterol converts to pregnenolone. The amount of aldosterone present determines the conversion rate of cholesterol to pregnenolone. This fact about the steroid manufacture rate remains true for the adrenals, testicles and ovaries. In the ovary, testicle and adrenal DNA programs, adequate aldosterone needs to direct production of the enzyme cholesterol desmolase. This enzyme makes steroid production possible from cholesterol. Cholesterol desmolase the initial enzyme for steroid manufacture is also known as a side chain cleavage enzyme.

A high potassium diet ('real food') stimulates aldosterone release from the adrenal glands. Processed food diets prove deficient in potassium and pave the road for dysfunctional ovaries. This fact provides an example of steroid interdependence in order for ovary health to become possible. Along with aldosterone, the healthy ovary needs vitamins C, vitamins A and B complex for adequate steroid production capabilities. The mineral zinc allows steroid synthesis to occur at high levels.

The mineral content between the two types of diets shows drastic differences. Real food contains natural minerals in the proportions required for health. Processed foods contain drastically altered mineral content. Processing natural foods depletes the minerals, magnesium and potassium. Unhealthy amounts of sodium, calcium,

hydrogenated fats, and sugar are added to extend shelf life. This equals profits for the food industry. Processed food doesn't contain the proper ratio of the four minerals - magnesium, calcium, potassium, and sodium. Ovaries need the correct mineral ratio more than other body tissues to maintain the aldosterone message content (see below). Aldosterone message content determines the synthesis rates for all steroids within the adrenals and ovaries. Natural food diets facilitate an increased aldosterone level.

Owners who eat dead food diets become potassium and magnesium depleted at the onset of middle age. A deficiency of either of these creates the need for an increase in insulin to process the same amount of sugar. The greater the insulin level, the greater the fat maker message becomes. In addition, potassium and magnesium deficiencies lead to high blood pressure, irregular heart rate, diminished adrenal and ovary steroid production, diminished red blood cell flexibility and decreased cell voltages (a weakened cell force field). The 'why' and the 'how' of these facts are discussed in the appropriate sections throughout this manual. Proper mineral intake equals the healthful state.

Chapter 8

A Deeper Understanding of Steroid Tone and Pressure

Weight loss is not possible until steroid tone and pressure improves. Aldosterone – maintains steroid tone and pressure. Aldosterone: an additional steroid hormone secreted by the adrenal glands. The central importance of aldosterone sufficiency concerns that its amount determines the amount of other steroids possible. Rather than overwhelm the reader about this important steroid initially its central importance has been delayed until now.

The adrenal glands respond to a narrow control range of salt and water balance that is largely controlled by the amount of aldosterone message content (an adrenal made steroid). The adrenal cortex secretes aldosterone within the glomerulosa layer. Potassium level, stress, and angiotensin type two comprise the three determiners for the amount of aldosterone that the adrenal gland releases.

The adrenal secretion of aldosterone directs three main events directed by its instruction of various cell DNA programs:

1. Increased sodium retention relative to potassium removal from the body.
2. Increased cellular charge (increased power to the cell force field) of cardiac and brain cells.
3. Increased steroid biosynthesis rates within the adrenals and gonads.

All steroid manufacture rates within the body depend on adequate aldosterone levels.

There are shared roles that the adrenals and gonads have in relation to steroid production. The quality of the types of steroid production within both of these glands will determine steroid tone. The amount of production of each steroid type determines the steroid pressure.

However, these two glands differ in the unique steroid types that they each produce. Aldosterone manufacture takes place in the adrenal glands. The amount of aldosterone produced determines the manufacture rate of the other body steroids. When aldosterone production goes well this benefits steroid production rates. When aldosterone production goes poorly, there occurs the liability of decreased steroid levels. The real food diet discussed earlier provides the safest way to increase aldosterone levels without raising the blood pressure. A side bar is included for the doctor readers who need more information before they will consider that aldosterone levels determine steroid tone, pressure and hence weights loss efforts.

Physician's Sidebar

The secret: prescription medications that raise nitric oxide levels

The popular blood pressure lowering medication called angiotensin converting enzyme inhibitors increases the histamine-like content of the body, which powerfully lowers blood pressure. An additional effect concerns its ability to increase nitric oxide production that also contributes to its blood pressure lowering effect. ACE inhibitors also conserve potassium, a known blood pressure lowering mineral. Finally, ACE inhibitors diminish ACTH secretion from the pituitary during the stress response. ACTH delivers the message to the adrenal to release aldosterone, DHEA and cortisol. Cortisol release at high levels leads to fluid retention. The mainstream textbooks say very little about these other powerful mechanisms for lowering the blood pressure. Instead, they discuss in great detail, the blood pressure lowering effects as being the result of lowered angiotensin two levels. They go on about how Angiotensin two is the main reason that aldosterone levels elevate. They skip the fact that most owners, with normal adrenals, who eat a real foods diet, high in potassium but low in sodium content, have high aldosterone levels, normal blood pressure and adequate steroid synthesis rates.

It is really quite a shock for most physicians when they begin to see evidence that the touted mechanism for a drug's action is not always the only way that they have an effect on the body. The ACE inhibitors, prescribed for lower blood pressure, provide one such example. The drug literature focuses almost exclusively on the supposed powerful role that angiotensin two plays in tightening up the blood vessels directly and indirectly, by raising aldosterone levels. However, very little of this literature discusses the well-documented fact that inhibition of this very same enzyme raises the total body content of a histamine like substance, bradykinin. This fact explains why a dry nagging cough occurs as the number one side effect. In addition, bradykinin lowers blood pressure directly but the price paid, as a side effect, manifests as increased leakiness of the capillaries in areas like the lungs and kidney. Maybe there was a marketing problem if this mechanism was related to increased histamine like substance content within the body. No one will probably ever know for sure. Nonetheless, it proves instructive to see a possible bigger problem with other drugs in how the physician gets 'groomed' into thinking about how these drugs work.

Because these drugs increase bradykinin within the body they also increase nitric oxide production as well. Increased bradykinin provides a powerful stimulus for turning on the nitric oxide synthesizing enzymatic machinery within the endothelial cells lining the arteries. Bradykinin and histamine share the same receptors in the body. They also act in a similar manner. They contain similar message content. Once bradykinin becomes elevated it tends to stimulate the mast cells to release histamine, as well.

The other touted benefit of these angiotensin converting enzyme inhibitors (ACE inhibitors), concerns their documented benefit in the preservation of kidney function. To understand that this benefit constitutes both a circuitous and expensive solution in many cases one needs to recall four things. First, ACE inhibitors conserve body potassium and this has a known kidney protective effect on the tendency to become potassium deficient. This medication then lowers blood pressure despite the elevated sodium in the body by the less realized mechanism of increased bradykinin within the body. Second, when bradykinin increases its presence this also lowers blood pressure by being a powerful stimulus for nitric oxide production.

Third, if a given patient was correctly counseled about a real food diet instead of a processed food diet, before kidney damage occurs from a chronically low potassium intake, blood pressure medication in many of these cases would no longer be needed (see mineral table).

The fourth fact to understand about the consequences of decreased angiotensin two production rates concerns its effect on the adrenal glands. Rather than get lost in the inconsistent evidence that these ACE inhibitors have on aldosterone levels, through angiotensin two directly, it becomes more instructive to look at the consistent evidence. The evidence is consistent that decreased angiotensin two will directly correlate with a decreased ACTH output. A decreased ACTH output will decrease stimulation to the adrenal glands for their release of aldosterone, cortisol and DHEA. This little detail has powerful implications for another mechanism as to how these medications lower the blood pressure. It also has powerful implications as to why diseases like autoimmune disease are made worse when these owners take these types of medications. Owners are made worse with autoimmune disease because they already have a wounded adrenal system (see *The Body Heal, Adrenal chapter*).

The addition of an ACE inhibitor will only exaggerate the diminished adrenal function, which operates in these diseases. This also provides a clue as to why these same types of owners will be at increased risk for neutropenia and lymphocytosis (see *The Body Heals, Adrenal chapter*).

Now it is time to turn the analysis on its head. The biggest roadblock mentally concerns the realization that people who consume high potassium and magnesium diets relative to total sodium intake are also going to have a high aldosterone (an increased potassium intake will powerfully stimulate aldosterone release). However, in these cases a diuresis will ensue because total body sodium is not excessive. Stress (ACTH) alone will tend to raise aldosterone secretion and in the situation of a high sodium diet this becomes inappropriate. When this happens blood pressure will rise. The point to consider is that perhaps a little effort spent counseling an early hypertensive on how to change their mineral intake ratios by eating 'real foods' (see mineral chapter) would have some merit before condemning them to medication with all its side effects.

ACE inhibitors have been shown to improve a heart failure patient's clinical situation. What is not said is that histamine like substances have a powerful strengthening action on heart muscle. This effect is obviously one benefit of these medications. In addition, the increased histamine like content within the blood stream will reduce the effort of the heart for getting the blood pumped with each beat. Rather than educate heart failure patients about the possible benefits of increased histamine, for pennies a day, they are prescribed expensive ACE inhibitors. Of course before histamine could be prescribed the scientific community would have to be persuaded to perform a safety analysis. The unlikeliness for this to happen provides one more example of the consequences of both science and medicine being dictated by profit. These facts point to just one example where what is good for the medical economy gets taught and sensationalized while the simple solutions collect dust.

This being said does not imply that ACE inhibitors are not needed. However, maybe it would be more honest to educate doctors and patients about simple solutions even though they may substantially impact the need for chronic medication bills and side effects.

Nutritional Adequacy: Effects on Steroid Tone and Pressure

Many physical signs occur in people who eat processed foods. Puffy, bloated faces, sagging skin, and loss of muscular definition exemplify the nutritional deficiencies processed food causes. Fast food restaurants are an excellent place to make these clinical observations.

Before this weight loss manual, unless you were an Asian who imbibed in good quality ginseng, less androgen would be manufactured with each passing year. Less androgen manufacture leads to lower steroid tone and the ungraceful decent into old age.

Poor nutrition causes gonad and adrenal cells to have a decreased ability to free up cholesterol and move it inside the mitochondria. The mitochondrion contains the site within the gonads or adrenal glands cells where cholesterol converts into early steroid precursors. Cholesterol needs to move into the mitochondria for the first step of steroid manufacture to occur. Once cholesterol moves inside the mitochondria it begins to convert into the various steroid hormones. The end product within the mitochondria is pregnenolone, a necessary starting molecule from which to build all other steroids. After pregnenolone forms inside the mitochondria from cholesterol, it undergoes conversion to other steroids outside the mitochondria. This internal conversion from cholesterol to pregnenolone becomes more difficult as the owner ages. The fall in pregnenolone leads to the consequence of a drop off of all other body steroids. Diminished steroid production has deleterious effects on both steroid tone and pressure.

There are four reasons the conversion of cholesterol to pregnenolone develops a roadblock with age:
1. Aldosterone content within becomes diminished secondary to a processed food diet (see below) or prescription medication.

2. Necessary vitamins and cofactors for biosynthetic reactions of these conversions diminish.
3. Cholesterol deficiency within the adrenals and gonads
4. The common method of taking pregnenolone supplements has possible detrimental side effects.

This fourth point needs additional emphasis before proceeding with the remedies for the other three causes of diminished steroid biosynthesis. The fourth factor exposes many short sighted and half thought out approaches for steroid replacement. Pregnenolone supplementation can partially bypass the need for these mitochondria first step reactions. The side effect of taking pregnenolone as a supplement for waning steroid production concerns the fact that it contains cortisol-like message content of its own. Increased cortisol message content is warranted in certain clinical situations. However, in most situations, when it occurs within one's blood stream after oral ingestion, it will contribute to decreased overall steroid tone and pressure.

Steroid tone and pressure becomes negatively influenced when cortisol message content rises above a very low threshold. In the normal state, pregnenolone never enters the blood stream in increased amounts. It converts within the adrenals and gonads into other steroids. In contrast, taking pregnenolone by mouth means that its blood levels will rise above normal physiologic amounts and only a small fraction will make its way into the ailing gonad or adrenal. This means that the lion's share of pregnenolone delivers its cortisol like message content throughout the body.

To counter act these problems, pay attention to the first three nutritional factors of adrenal and gonad support. The first of the three nutritional factors affecting the health of these glands involves the owner's history of diet preference. This determines the aldosterone level of most owners. Aldosterone levels are affected by either a 'real food' diet or a 'processed food' diet. As aging proceeds, eating processed foods leads to an eventual depletion of one's mineral balance. Around middle age owner's bodies become deficient in mineral balance through the chronic ingestion of processed foods. Processed foods have the wrong mineral proportions (see mineral chart for evidence). The right mineral proportions are necessary for high steroid tone. Real foods (unprocessed and natural) will have a much higher proportion of potassium and magnesium content. Real food's sodium content is lower than processed food. The opposite situation of mineral balance exists in owners who arrive at middle age with a history of a preference for processed foods (dead food). These food types contain diminished potassium and magnesium content from the act of food processing. Processed foods usually have high amounts of sodium added as well. Many middle-aged owners, who eat processed foods, prematurely suffer the consequences of a decline in steroid tone and pressure.

The central steroid defect from a chronic 'dead food' diet arises from the diminished aldosterone production required for normal blood pressure to be maintained. Some bodies still try to produce appropriate aldosterone in this chronic situation of a dead food diet, but high blood pressure predictably results. The dead food diet contains high sodium and low potassium. The human body was designed to ingest the opposite of the reverse mineral content, which the processed food diet provides. Sodium retention beyond healthful amounts causes fluid retention, which causes elevated blood pressure. The other group of owners has bodies that sense inappropriate sodium content and hence they reduce aldosterone levels in order that the blood pressure stays normal. The price paid by these owners' bodies results in a diminished steroid synthesis rate. Steroid synthesis rates depend on ample aldosterone levels.

The owner who takes an active role in the procurement of a real food diet (exiting the torture chamber chapter) has an advantage. The 'dead food' (processed) diet accelerates the path to an old body in two major ways. First, diminished aldosterone or elevated blood pressures are the choices given to the body that feeds on dead food. When a body chooses the lower aldosterone route, steroid tone and pressure suffers and the owner ages faster. Conversely, when a body chooses to maintain steroid tone by keeping aldosterone elevated in the presence of a processed food diet, the blood pressure elevates. When the blood pressure elevates the owner gets older due to a failure to honor a central principle of health, avoiding hardening processes. Elevated blood pressure ages the body by its ability to increase hardening processes within the blood vessels.

The second reason steroid production rates begin to fall with advancing age results from a deficiency in vitamins and cofactors. Vitamins and cofactors are necessary for the manufacture of many different steroids. These additional nutritionally dependent factors affect steroid tone. Enzymatic machinery must be in good working order to allow efficient steroid biosynthesis. Individual enzymes (cellular machines) need specific trace minerals and cofactors in order to perform in the creation of the different steroids. Only when steroids are manufactured at appropriate rates can steroid tone and pressure remain high.

An analogy to this is the automobile that has a full tank of gas, but all the oil has been drained out. In addition, the spark plugs and carburetor are missing. Missing car components are like the cofactor and vitamins steroid producing enzymes (machines) needed to make the necessary steroids. When a cofactor deficiency occurs, steroid production diminishes. When steroid production diminishes steroid tone and pressure falls.

A partial list of vitamins required for the manufacture of steroids is: vitamin A, pantothenic acid, folate, most of the B vitamins, and vitamin C. Under healthful conditions, the adrenal gland has the highest tissue concentration of vitamin C in the body. This fact might provide a clue as to why vitamin C is so important in the owners continued survival during stress and in the prevention of illness. The stress of surviving an illness requires an increase in steroid production. The need for vitamin C goes up with stress and illness.

The accessories, in the car engine analogy, lend themselves to understanding the reason that dietary attention and discretion become so important as owners navigate their search for health and happiness. Youth can survive for a while without the critical nutrients needed by cellular machinery. Youthfulness that endures cannot. 'Father time' is there to observe old cellular parts struggling to work beyond their normal life span without the necessary molecular replacement parts. The body can only acquire new molecular parts through dietary intake of the molecules needed (see digestion chapter). Food processing destroys important nutrients required for the production of enzymes. There are at least five B vitamins contained in whole grains that decrease by half within one week of grinding it into flour (half life = time for half to be gone). Oxidation of these unstable and bulky molecules occurs after the grinding process. Vitamins are positioned precisely within the seeds molecular architecture to confer stability. When grains or seeds are ground into flour this process destroys the protection provided by the seed's architectural framework. The grinding process exposes these unstable molecules to the oxidative forces and thus diminishes their nutritional value.

Oxidation takes place in vitamin-fortified foods exposed to heat, air and sunlight for the same reasons. Just because manufactures add vitamins to the box or can doesn't mean they are still intact when the owner ingests it. In contrast a real food diet still contains vitamins and minerals that the creator intended.

Vitamin supplements are more stable, but absorption characteristics differ widely from brand to brand. Vitamins and vitamin-fortified foods are deficient in some of the vitamins that are removed when food is processed. Folate, pantothenic acid, and lipoic acid are especially susceptible during food processing. The deficiency problem is compounded by the fact that many vitamin fortified foods and supplements are deficient in these three vitamins.

When one of these is missing it seems like a malicious conspiracy exists within the food industry that makes owners weak and old.

The weakest link in a chemical reaction sequence will stop a body process. In a pantothenic acid deficiency, carbohydrate and fat combustion greatly curtail. With a lipoic acid deficiency, carbohydrate only converts to lactic acid that builds up in the tissues and evidences itself as sore and achy muscles. Adrenal and gonad tissues need tremendous amounts of fuel for energy.

Steroid tone and pressure diminishes when these vitamins become deficient.

Additional trace minerals are needed within the adrenals and gonads for maintenance of steroid tone and pressure. Minerals like zinc and magnesium need adequate stomach acid for absorption. Many western owners are prescribed medicines that block acid production. These owners could become deficient in many of these critically needed minerals despite a healthy diet.

A diet containing sufficient organic and unprocessed whole grains, vegetables, eggs, fish, meats, nuts and fruit promote steroid tone and pressure. A 'real food' diet promotes steroid tone and pressure because it contains the proper proportions of minerals and vitamins to promote the health of steroid producing glands as well as other body tissues.

Many plants also contain antioxidant (anti-rust) molecules that confer particular benefits to specific tissues. Examples include lycopene found in tomatoes for prostate health, bioflavonoids found in berries for blood vessel health, acanthocyanin found in bilberry for retina health, silymarin found in the milk thistle for liver health.

Carelessness in how these foods are processed can destroy much of the anti-oxidant content and their potential benefits. The lower the antioxidants in the diet, the more need for anabolic steroids to direct increased repair. Increased repair rates become necessary when diminished protection from the oxidants occurs. When oxidant damage occurs, there becomes more need to elevate steroid tone and pressure to compensate for the increased injury rate. Increased steroid tone and pressure compensates for the increased injury rate because when steroid tone remains high, anabolism remains high. In order to derive the intended benefits of sufficient steroid tone and pressure, purchase only quality supplements or real foods that comes from fresh organic sources.

Environmental Toxins: Effects on Steroid Tone and Pressure

Many toxins exert their effect on adrenal and gonad health. Environmental effects on steroid tone (hormone mimics subsection) increase the estrogen-like message content in an owner's body. Increased estrogen message content creates problems in anabolic balance and in estrogen responsive tissues that include men's prostates. Many of these mimics and toxins prove far reaching in their ability to disrupt the ideal natural hormone message content. When natural hormone message content imbalances, so does steroid tone. Effective measures limit exposure to environmental toxins that diminish steroid tone. Limiting exposure to these toxins can restore the message content provided by anabolic steroids.

Several lifestyle practices limit the amount of toxins that reach the steroid producing glands. First, limit exposure to hormone mimics. Consume organically grown food whenever possible. The food supply provides the mechanism for how most of the hormone mimics invade an owner's body (see hormone mimics subsection). Clean water consumption avoids many of the hormone mimics. Stay away from golf courses because of the heavy use of hormone mimic-like molecules. Education about toxins in the home is important. Avoid the counter-productive stress of being paranoid; just try to reduce chemical exposure within reason. Ideally, with increased public awareness about the consequences of continued industrial, agricultural, and food distributors practices, which allow this problem to perpetuate, things will change for the better.

Emotional Energy: Effects on Steroid Tone and Pressure

An owner's overall emotional quality operating in his/her life has a powerful effect on steroid tone and pressure. Positive emotional experience communicates to steroid producing glands to increase the steroid tone mixture output. Negative emotions influence steroid glands to believe survival is the issue. When the body perceives a survival threat, either real or imagined, stress steroids increase and rejuvenation steroids decrease. When chronic negative emotions occur, this mechanism lowers steroid tone.

In many respects, America is a stressful place to live. Deadlines, job insecurities, modern life complexities, noise pollution, environmental pollution, and electromagnetic pollution take their toll on the human body. The body experiences these emotional stresses energetically by the informational substances it secretes. The stress response informational substances include cortisol. When cortisol levels increase the relative proportion of its message content among other steroids increases as well. This proportional change will lower steroid tone.

There are specific dynamics in the way cortisol opposes androgens and the energies that direct this process. The dominant emotional energy determines how the body spends available energy. The different emotional energies determine, in part, the hormone messages that secrete. In this way, the different types of emotions have a profound effect on energy usage patterns.

Studies have shown that teammate's testosterone increases immediately before a competition and remains elevated during the event. Studies also show that the teammate's testosterone will remain elevated when the game is over, only if they win. These facts provide the contrast between the negative emotions of losing and the positive emotions of winning. The outcome of the event not only affects the quality of emotions, but the quality of steroid tone and pressure. Studies have also shown that chronically depressed owners have lower DHEA levels than aged matched controls that are not depressed.

Quality of the emotions effects on overall steroid tone and pressure can be understood in a general way. The adrenals have the choice of making more DHEA or more cortisol. Positive emotional experiences usually allow DHEA levels to remain adequate. Positive emotions translate into the body as a low stress situation. When the body perceives a low stress situation, more energy becomes available for rejuvenation activities. When the body perceives the survival threat of negative emotions, more cortisol becomes needed. The body has limited energy. Hormones direct energy expenditure. The stress response channels energy expenditure by the types of hormones that are summoned.

When negative emotions cause stress hormones to increase, body energy directs into catabolic pathways. Catabolic pathways consume body structure. The survival response involves the  inhibition of rejuvenation activities and increases fuel availability within the blood stream. Increases in fuel availability brought on by stress are a primitive design feature. Modern stress differs from primitive stress, which usually involved a physical challenge. Physical challenges need extra fuel to maximize strength. Modern mental stress requires no physical response. The result of this design feature, in a setting of chronic negative emotions, results in the increased catabolic direction of energy and decreased anabolic direction of energy. The cortisol message content increases and the androgen message content decreases.

DHEA is one of the body's main androgens. All androgens convey a general cellular message that says to invest adequate cellular energy into rejuvenation and infrastructure investment. Cortisol, because it is catabolic, directs cells to put a hold on cellular rejuvenation and infrastructure investment. In addition catabolic hormones, like cortisol, direct the dismantling of body structures, like protein, to provide fuel to the blood stream. Negative emotions direct a powerful message that channels energy into catabolism by increasing cortisol predominance. Only a normal cortisol level conveys a harmonious and less dramatic message. The resultant high cortisol excretion rates, when the body perceives a survival threat, direct energy toward survival pathways. Survival pathways only become necessary for mounting a strong physical response at the expense of cellular maintenance and rejuvenation. Emotional energy has a powerful effect on steroid tone and pressure.

A sufficient steroid tone and pressure is fundamental for weight loss.

Excess cortisol has the ability to direct energy away from youth maintenance activities. The message content from high cortisol levels directs energy toward survival. Some scientific circles have called the chronic elevation of cortisol the 'death hormone.' This title refers to chronic deference of cellular maintenance activities when cortisol becomes chronically elevated. This accelerates wear and tear changes because there becomes less anabolic message content. One of the deterrents to wear and tear changes (old age) involves having adequate message content to rejuvenate. Message content to rejuvenate comes from the anabolic steroids. When the anabolic steroids balance with the catabolic steroids, steroid tone and pressure maintain and health remains more likely.

A subtle but important negative feature of elevated cortisol levels now arises. Healthy people with low stress levels obtain their cell fuel between meals from the release of growth hormone. Growth hormone release causes the simultaneous release of IGF-1, sugar and fat into the blood stream. Even more importantly, concerns the fact that adequate growth hormone release also protects the body protein stores from catabolism. Remember that the other fuel releasing hormones, glucagon, epinephrine and cortisol all increase the consumption rate of body protein stores. Proteins comprise the metabolically active component of body tissue (enzymes, mineral pumps, muscle and organs). The greater the metabolic rate the more calories burned. When something goes wrong in the normal chain of events of growth hormone release followed by normal amounts of IGF-1 release, body protein is lost.

Here lies an often-unrecognized vicious cycle within the torture chamber on the path to an old body. Unhealthy people are trapped within the torture chamber on the path to old age because no one counsels them on how to increase their growth hormone, IGF-1 and hence protect their body's protein content. Bodies that rely on cortisol, epinephrine and glucagon between meals to keep the blood fuel elevated will lose protein and gain body fat every time. Deficient growth

hormone or IGF-1 secretion leads to an obligatory elevation of hormones like cortisol between meals or when exercising. Cortisol contains catabolic message content that makes protein combustion accelerate for fuel needs. It is the protein component of body mass that allows calories to burn (muscles, organs, enzymes and mineral pumps). The less protein within a body, the less need for caloric expenditure. The fewer calories needed, the quicker one gains fat.

Making the obesity creating torture chamber worse concerns the fact that a body that feeds itself between meals with directions from hormones like cortisol will also need more insulin. More insulin becomes necessary because the fall off of growth hormone also causes a fall off in IGF-1. Adequate IGF-1 greatly reduces the need for insulin. IGF-1 and insulin are the only fuel nozzle hormones within the body. When one fuel nozzle hormone decreases the other needs to increase so that there are enough fuel nozzles to fill all the cell fuel tanks. Remember, only insulin contains the fat maker message, which greatly exaggerates within the liver because of its site of release from the pancreas into the portal vein. The portal vein leads straight into the liver. This means the liver will always receive the highest concentration of insulin whatever the amount of its release.

As a general rule, the faster the metabolic rate in body tissue, the more vulnerable it becomes to deferred maintenance activities. High metabolic rate tissues, like brain and heart tissue, consequently become more vulnerable when decreased steroid tone and pressure occur. The overall quality of the mixture determines high or low steroid tone. High steroid tone quantifies the tendency for percentages of different steroids in the blood stream to approach optimal. High steroid pressure means that adequate amounts of the individual steroids reach all parts of the body. Inadequate steroid tone and pressure will show itself clinically in the body periphery first: the joints, the skin and the bones.

High steroid tone and pressure directs the body to invest appropriate energy in the repair of the wear and tear associated with changes inherent in life's activities. High steroid tone directs appropriate rest intervals and is synonymous with the balance of message content between anabolic and catabolic steroid hormones. Optimal steroid tone needs to occur if youthfulness and healing are to occur. The quality of emotions has a powerful effect on the consequent steroid tone.

That emotions effect steroid tone is an essential concept to combat the ravages of life and emerge as an owner who ages gracefully and lives life fully. The fact that positive emotions promote high steroid tone means there is an increase of appropriate levels of testosterone, DHEA, androstenedione, and progesterone relative to the total amounts of cortisol-like substances within the blood stream. The balance between opposing steroids makes the difference.

The healthful state of the human body requires re-building activities balanced by rest periods. Extremes in the building up activities and the owner become the stiff body builder and/or the aggressive personality types. When energy channels to the extremes of a body builder physique or the aggressive personality type, too much unhealthy anabolism takes place. Cortisol induces appropriated cellular rest periods in the healthy state. Balance needs to occur between rest and adequate periods of androgen secretion. Emotional energy either facilitates or upsets this balance.

The physical stress of vigorous exercise causes temporary elevation of cortisol that directs energy toward maximal strength activities. This means cellular maintenance decreases. The difference between exercise stress and emotional stress concerns the fact that physical exercise increases testosterone as well as cortisol when emotions allow it. In addition, when an owner remains healthy, exercise increases growth hormone secretion and hence protein content has some protection. When the emotions allow it, testosterone remains elevated after exercising and cortisol levels decrease. During the stress of physical exercise there becomes no need for extra insulin because the increased blood sugar, directed by cortisol, is appropriately consumed during physical exercise. Insulin increases become unnecessary because the exercising cells of healthy owners uptake their nutrition from the adequate IGF-1 released from growth hormones presence. In marked contrast with emotional stress, in unhealthy owners, only cortisol secretes. The message of cortisol completely differs in the absence of the testosterone message. The message difference results more from the increased insulin required and the lack of testosterone when mental stress occurs. Making matters worse is the fact that the elevation of blood fuel, which cortisol directs in the stress response, inhibits growth hormone release. Growth hormone release occurs in response to a falling blood fuel level.

The stress response, with its cortisol directed rise in blood fuel, circumvents the need for growth hormone and has two health consequences when it becomes the norm. First, without adequate growth hormone secretion there will be a loss of body protein content. Because proteins comprise the metabolically active component of body tissue, metabolism slows. Secondly, insulin needs increase because when growth hormone secretions fall off, IGF-1 levels drop off as well. More insulin is needed to make up the deficit of fuel nozzle hormones. Fuel nozzle hormones feed the body cells the nutrition in the blood stream. The insulin fuel nozzle hormone preferentially fills up the liver and fat cells with sugar. Liver and fat cells turn the excess of sugar into body fat and cholesterol. In contrast, IGF-1, the other fuel nozzle hormone, occurs at blood levels greater than 100 times those of insulin, in the healthy owner. In this way the healthy owner's other body cells have a way to procure blood fuel without the fat maker message of insulin.

Part of the chronic stress response involves soaring cortisol secretion relative to other steroids. Chronic stressful emotions raise cortisol levels relative to other androgens. This accelerates wear and tear within cells because cortisol has the advantage of directing cellular energy in chronic stress or emergency situations. Body response to an emergency situation is always the same. The body directs energy out of rejuvenation and increases fuel availability from body structures (protein, fat, and carbohydrate).

The more metabolically active the body tissue, the more vulnerable it becomes to deferred maintenance activities. The brain has one of the fastest metabolic rates within the body. Therefore, chronic emotional stress has the potential to contribute to the rate of brain aging quickly. The brain ages more quickly because energy chronically directs into survival pathways. When survival pathways activate, rejuvenation (repair and maintenance) activities inactivate. These changes become physically manifested in diminished cell membrane integrity, aging intracellular factories, and enzymatic machinery. A tendency for oxidized fats and proteins to build up inside cells also occurs. As oxidized fats accumulate, they condense into waste known as lipofuscin. Clinically, the aging brain manifests in a slower reaction time and progressive memory impairment for new events and concepts.

The dominant hormone message often provides a reflection of the dominant emotional energy. Positive emotions of love, forgiveness, joy, hope, and contentment convey to the body that all is well. The body naturally produces steroids consistent with the emotional state. Conversely, negative emotions of hate, anger, fear, and sadness direct an increase in the production of stress hormones. Beyond the effects of emotions on steroid tone and pressure are factors that increase rejuvenation-type steroids. Scientists call such factors anabolic secretogogues.

Secretogogues Enhance Steroid Tone and Pressure

Secretogogues enhance the output of the rejuvenation message content contained within the steroid class called androgens. Androgen message content increases when secretogogues increase. Examples of secretogogues are regular exercise and ingesting certain plants like panax ginseng.

Counterbalances to stress are the practices that promote secretion of strength and rejuvenation steroid hormones (DHEA, testosterone, and progesterone). Secretogogues promote anabolic glandular secretion. Given adequate nutritional support and adequate adrenal and gonad health, certain practices promote optimal secretion of body building steroids and promote an elevated steroid tone. Without some of the three criteria of adequate nutritional support, adrenal and gonad health being present there cannot be an adequate response to a secretogogue. Secretogogues can't pull strength hormones out of a dying gland. The shriveling gland affects other functions.

An analogy for this involves the similarities between factories and cells. Old gonads and adrenals are like an order arriving at a factory and the condition of the factory is one of disrepair. The disrepair is evidenced by aged factory support structures (organelles) and aged machinery (enzymes) that is worn out and falling apart. Very little, if anything, can be done to fill the order for hormone creation. Before the factory can respond to the order, it needs to be remodeled and upgraded in its infrastructure and machinery. The adrenal and gonad 'factories' are similar in that they need the proper infrastructure investment provided by adequate molecular replacement parts and the right hormones (message content) for repair directions. Sedentary owners send the wrong message content (hormones) and their 'factories' tend toward disrepair. This fact provides the root of the saying, 'use it or lose it'. Exercise is a secretogogue directing cells to improved message content and invest in rejuvenation.

Owners with poor steroid tone and pressure hate to exercise. This fact results from long standing habits that produced an attitude. Part of the problem lies in the fact that sedentary owners have a diminished secretogogue influence in their lives. A significant biochemical component for this aversion exists.

Poor steroid tone and pressure are predictable when the physical condition deteriorates. Out of shape owners predictably have steroid mixtures that reflect their poor physical exercise habits. Conversely, a well-trained athlete possesses sufficient types and amounts of steroids that reflect high steroid tone and pressure. Highly trained and physically fit cells of an athlete invest maximum cellular energy in maintenance and rejuvenation. Their cells receive the proper message to invest in rejuvenation. To the highly trained athlete, regular exercise acts as the secretogogue stimulant for anabolic steroid production.

Chronic stress of modern life makes matters worse by encouraging cells to invest energy foolishly, which accelerates wear and tear on the body. Stress reroutes life-sustaining energy into survival mode. When poor exercise habits subtract out of the daily routine, deterioration becomes more aggressive in assaulting the body form. The higher the stress in life, the more need to counter balance with secretogogues. Exercise proves a reliable secretogogue. Owners in the middle of modern life stress need to exercise.

Unhealthy owners need to know that their body wants to heal. Their body needs the correct informational message to do so.

Higher steroid tone carries the information needed to heal the body. Optimal steroid tone doesn't return over night. It is a gradual process just like getting unhealthy is a gradual process. Understanding that with the passage of time, vigor and vitality will increase as the owner persists in improving the quality of informational messages that his/her cells receive. The secretogogues like exercise help owners begin healing. Their cells begin to receive improved information from higher steroid tone that exercise provides. The secretogogues increase the amount of energy available for cellular rejuvenation and infrastructure redevelopment. These restorative activities increase because steroid tone increases.

Owners committed to their recovery program notice positive changes every time they look in the mirror and from the improvement in the way that they feel as the months go by. Recovery doesn't have to be expensive or complicated. Recovery can start today with a commitment to understanding the nutritional needs and the emotional environment that gives and takes energy from the owners life. The body begins to heal with improved nutritional quality and through a daily walk, stretch and breath exercises.

The ability of exercise to increase steroid tone and pressure crosses over into its effect on emotional energy as well. Positive emotions confer an energetic rhythm to cells by facilitating improved quality of informational substances that course through the blood stream. The emotional energy of happiness, joy, laughter, self-love, and forgiveness are all ways to increase the rate of healing. Exercise adds positive emotional energy to an owner's life. This is effective beyond its secretogogue influence. An owner is asked to remember the last time they felt happiness and what it felt like. This is instructive as far as remembering what the cells felt via the informational substances. Happiness energy directs reflective release of the appropriate informational substance (hormone) that communicates to the cells that all is well. This message allows the cell to harmoniously interact with maximum efficiency and aliveness.

A deeper understanding of why adequate aldosterone is important for weight loss success

Aldosterone can be thought of as a misunderstood steroid hormone. Currently a narrowness of thought limits an appreciation for the impact that aldosterone has on certain cells. Many physicians think in terms of water and salt balance only when considering aldosterone. Aldosterone influences certain cell types in their ability to increase cell membrane charge. Maintenance of an optimal cell charge only takes place when there is adequate instruction from aldosterone and thyroid hormones. These two hormones instruct certain DNA (genes) of cells to activate. These specific activations lead to an increase in cell membrane mineral pumps synthesis rates that increase the force field (cell membrane charge). In the maximal function of cardiac cells and nerve cells this is particularly important. The stronger the force field created by a cell type, the more work it is capable of performing. In addition, a more powerful cell electrical charge gives added protection from inappropriate penetration of harmful ions (excess calcium). This is important for the continued operation of a nerve cell.

One of the indicators of aging within concerns diminished cellular charge. When the overall one hundred trillion body cells have diminished cellular charge, metabolism slows. **In fact, the maintenance of cell charge consumes the majority of calories in the resting state.** Owners with a diminished cell charge burn fewer calories and therefore gain weight more readily. Lastly, aldosterone message content has additional roles in the instructional activation of gonad and adrenal DNA (genes). When the DNA programs of these glands activate in this way, the steroid manufacture rate increases.

As discussed earlier, certain medications, called angiotensin converting enzyme (ACE) inhibitors, supposedly lower blood pressure by poisoning the ability of the adrenal gland to release aldosterone. Currently a debate exists concerning the exact mechanism of how these prescription drugs work. Aldosterone diminished release is the mechanism physicians are given (see earlier discussion).

Blood pressure lowers through several others but under reported mechanisms (see previous discussion). For the purpose of the current discussion, only the aldosterone lowering mechanism is pertinent because of its secret side effect.

Very few doctors are educated about the importance of aldosterone in stimulating the adrenals and gonads into steroid production. A lower aldosterone message content to the heart and nerve cells also diminishes the ability of these cells to increase their force fields to the strength required for maximum function. Diminished force fields within nerve and heart cells lead to an inability for these cells to perform maximally.

There are circumstances where aldosterone production must be suppressed. When physicians become aware of the importance of aldosterone, they will understand the appropriate times and methods for accomplishing this. They will also avoid reducing aldosterone production when suppression is inappropriate.

This narrowness of learning that occurs in medical educations provides one example of how well informed patients can help curious physicians understand what science has revealed. Alternative methods for natural ways to heal high blood pressure are more fully discussed in *The Body Heals*. The current medical system controls the behavior of physicians through a mixture of three things. Acceleration in busy work takes away from a doctor's time and desire for new learning. Second, people that become 'certified experts' are the ones who learn early on to keep their mouth shut when they stumble upon scientific inconsistencies while being indoctrinated with the official view of medical treatment strategies. Third, the medical legal rule of the 'standard of care' must not be violated. Any deviation from the sanctified approach will alienate a physician into a legal battle with the complex. The medical industrial's viewpoint has profit to consider.

This being said, never under estimate the power contained in even one man's life when he lives it properly. Like ripples in a pond spreading forever outward, so it is when at the grass roots, people begin to create ways for their physicians to again practice and learn different ways that lead to healing.

Chapter 9

Obesity Has a Component of Causality in Poor Nutrition

Many owners stay fat because they have not been helped to understand what their body's need. Bodies receiving the nutrition they need stop craving the next feeding event. A major component of obesity perpetration resides in **the hungry cell syndrome**. The hungry cell syndrome results when the wrong hormones release and shunt fuel inappropriately into storage (fat). Owners that understand this fact can maintain a focus on ways to realign their hormones message content (see earlier glandular chapters discussion). Similarly, insight into what good nutrition entails and how it happens further facilitates a successful weight loss effort. This section concerns the understanding of what food is and how the body assimilates it into the body structure. In the following chapter, a trip down the digestive tube allows a greater awareness of the miracles that occur. A basic nutritional knowledge empowers the overweight owner to successfully shed fat and rebuild body structure.

The prevention of body breakdown occurs because only functional digestive tracts can maximize the absorption of nutrients. The owner needs to make good nutritional decisions or there will be insufficient building blocks for the digestive tract to absorb.

Reactions of life use up molecular parts quickly. A continual need for new sources of quality molecular parts exists throughout life. Molecular replacement parts serve in making body structure, fuel, maintaining cellular electrical charge, hormones, and as part of enzyme machine activities.

Structures that are alive, cells and tissues, require a constant supply of quality molecular replacement parts. The process of life degrades molecules. This chapter concerns how these molecular replacement parts are obtained and assimilated. When replacement parts assimilate, the rejuvenation program continues.

Inadequate molecular replacement parts leads to an old and flabby body. The pursuit of longevity concerns how to obtain and assimilate quality molecular replacement parts. Quality parts are necessary to prevent the old body manifestation that results from the accumulation of worn out parts. Bodies with worn out parts have hungry cells that desire new molecular replacement parts. The obvious consequences of inadequate molecular replacement parts or a deficiency in quality fails to be comprehended by many owners. Often this disconnection results from scientific communication being explained in an abstract manner.

Each cell type needs a continuous supply of specific replacement molecular parts. These molecular replacement parts are needed in specific amounts and sufficient quality to maintain healthy cell function. The processed food diet is deficient in regenerative replacement parts. Deficiency, excess or inferior quality interferes with health. Obesity describes one manifestation of diminished health. Obesity results, in part, from the hungry cell syndrome.

When the digestive tract fails, health consequences ensue such as:

- Cellular trash accumulation
- Loss of cellular integrity (old and worn out molecular parts)
- Deficient molecular parts needed to manufacture adequate informational substances (hormones and neurotransmitters)
- Inadequate cellular charge (mineral imbalance)
- A progressive deficiency in the ability to maintain cellular energy requirements (vitamin deficiencies, rust and poor quality fuel intake)

The Body is Analogous to a Temple that is Alive

The ongoing assimilation of molecular replacement parts is similar to the non-living building world. In each case fundamental building blocks, each with specific roles create the overall whole (wire, bricks, two by fours, windows, light sockets, trusses, etc.). These components have the potential, when assembled correctly and proportionally, to become an elaborate temple. Multiply the architectural complexity of the ongoing body molecular parts replacement program. Throughout life this replacement program proves necessary for healthy structure and function.

The body can be thought of as a dynamic, living temple that will, in time, wear out its constituent molecular parts as it pulsates with the energies of life. The wear and tear creates accumulation of defective molecular structural components and enzyme machines. The body temple requires precise maintenance of internal structure and energy flow parameters. It can only obtain these parts from other temples (living things). These other temples must be precisely dismantled or rubbish instead of replacement parts results. The precise dismantling of these temples describes the process referred to as digestion.

Healthy owners have digestive systems that assimilate their continuously needed replacement parts into their body. Availability of sufficient parts results from precise dismantling of other temples in the digestive tract. The absorption of these molecular building blocks and fuel sources (to power the 'intratemple electronics') occurs in a rhythmic fashion. After absorption of these parts, they reassemble into complex components (cell structure or enzymes) or combust as fuel in the cellular power plants (mitochondria). The reactions of life diminish when the digestive tract malfunctions in its ability to precisely dismantle a molecular building block. The lack of replacement parts leads to many chronic degenerative diseases. Chronic degenerative diseases have a component of causality in the hungry cell syndrome. Obesity is a chronic degenerative disease.

The hungry cell syndrome results from:

- Poor nutritional choices
- Improper dismantling ability of critical structural components that the cells need to replace their worn out structural components or old enzyme machines
- Poor quality communication between digestive cells secondary to imbalanced hormone tone
- Injury to various digestive chambers and structures

Hungry body cells turn up the appetite centers volume. These owners are constantly in the torture chamber of desire for the next feeding event. The torture chamber always wins until they are taught how to feed their cells with what they need. Well-fed body cells do not cry out for more food. The processed food diet perpetrates the hungry cell syndrome. The hungry cell syndrome propagates when molecular replacement parts become deficient.

Digestive Tracts and Molecular Supply Lines

All too often the nutritional integrity of cells is superficially addressed. Many owners age at an accelerated rate unnecessarily. The disassembly of protein, carbohydrate, and fats relies on numerous digestive juices. These juices must occur in a specific sequence and be of a sufficient amount. Protein disassembly provides a prototypical example of the three different food groups that can facilitate comprehension of problems with the dismantling process - digestion.

Proteins are comprised of twenty different molecular shapes (amino acids linked end to end). Each amino acid type has a different shape and contains unique chemical qualities. The order of sequence of amino acid connection and the proportions of different amino acids determine the resulting properties of a protein.

It helps to visualize twenty different miniature animals as analogous to the twenty different amino acids that make up all proteins. Different animals occur in various sequences and numbers as they line up. The average number of amino acids strung together in a protein equals 500. This allows a mental picture of the average protein molecule. This chain would have so many tigers, so many elephants, etc. The specific order of these shapes and the amount of each type confers a unique chain (twenty different building block shapes, corresponding to twenty different amino acids that link together to form a protein). In order to build this chain, the body can only obtain the different lions, tigers, and bears from properly dismantled (digested) other chains (protein meals). In order to build new necklaces, a tiger separates as a whole tiger and not a half of a tiger and half of an elephant. If this occurs, the building block potential of the different shapes reduces to rubbish.

The body can produce only twelve of the different amino acid types. It cannot manufacture the other eight. When the body manufactures its own amino acids it depletes essential ones that must be obtained in the diet. The best situation occurs when an owner consumes proper proportions and amounts of all twenty amino acid types.

The healthy digestive tract, when presented with these chains (proteins), starts methodically dismantling them. The proteins must be precisely disassembled or useful replacement parts cannot be absorbed. When amino acids absorb, they can be used for building new proteins (chains). The unique order and amounts of different amino acids confer specific qualities and abilities for that protein. Some proteins contain many sulfurous amino acids. Other proteins have extra nitrogen. Still other proteins have extra acid content. The overall order and content determines protein function.

Some proteins are used for structural integrity (cell membrane, organelles, hair, bone, ligaments, etc.). Some serve as transport vehicles in the bloodstream for critical minerals, steroids, and gases (calcium, testosterone, and oxygen respectively as examples). Other proteins function as cellular machines (enzymes). Still others form into the many different types of informational substances (insulin, glucagon, and IGF-1).

The body requires a continuous supply of quality building blocks (amino acids) in order to maintain availability of different protein types. Amino acid building blocks need to be structurally intact when absorbed. The digestive process fails unless the consumed protein meals dismantle precisely into their component amino acids.

Molecular Replacement Parts Require Precision Dismantling

The public is led to believe digestion is analogous to food that grinds up in a blender. The perpetrated story implies that this slurry is then sucked into the bloodstream and divided among the different body cells for use. The temple analogy can expose this gross simplification.

If a building explodes, rubbish results because there is little salvage ability in the way of reusable structural components (intact doors, windows, trusses, flooring, appliances, etc.). In order to have any reusable structural components derived from a temple, there needs to be a precise dismantling process. There can be no reusable building components without precision dismantling occurring first. The human body digestive process similarly needs to be orderly and precise in its ability to dismantle the various food building blocks, nutrients, vitamins, and chemical reaction facilitators (cofactors).

Shrink down to the size of a molecular glass ship. In this glass ship take an imaginary trip through the digestive tube. The trip down the digestive tract from mouth to anus involves passage through multiple chambers. Each chamber concerns a specific digestive process that needs to occur before movement to the next chamber becomes appropriate. These processes occur all along a healthy digestive tube.

The digestion that follows each meal needs to be an orderly and precise process. This process dismantles the food consumed into building block molecular components. Next, these molecular components absorb into the body at specific sites along the tube. Once these different molecular replacement parts absorb, they eventually secrete into the blood stream. There they distribute to different body cells and reassemble into new temple components or burn up as fuel. Disease and old age results when the supply of these necessary replacement molecular parts becomes scarce.

Only with adequate molecular supply lines, adequate workforce, optimal workforce environment, integrated communication with the salvage team and sufficient enzymatic machines will there be sufficient cellular rejuvenation. This leads to new cellular structural repairs, new transport vehicles, new cellular factories, new informational substances, and adequate cellular trash removal.

Without adequate cellular rejuvenation all the cells are forced to make due with old cellular machines (enzymes). Also, cellular trash accumulates. The cellular factory components age because they work long after their useful life spans.

The Task at Hand for the Digestive Tube

The accomplishments of a healthy digestive system after a meal is best understood when the different components of that meal are isolated.   These isolated components need to be discussed in regards to their unique contribution and their particular dilemma before body assimilation occurs. The dismantling of protein was already reviewed above. Some of the other meal components disassembly are discussed below. A trip down the tube in the glass ship follows the basics of digestion review.

How to Dismantle Carbohydrates

Carbohydrates break down into sugar and comprise the quickest food group to be absorbed. Almost 100% of sugar absorbs within a healthy owners digestive tract at the rate of 120 grams an hour (about 500 calories). Fat absorbs the slowest and protein assimilates a little quicker than fat. Hunger returns more quickly following carbohydrate meals. One of the reasons concerns the fact that this fuel type is removed quickly from the digestive tube. An empty digestive tube leads to a strong desire to eat. Carbohydrate intake also obligates the insulin hormone's release. Exaggerated insulin release becomes necessary in overweight owners because some other glandular process has failed. For example, insulin processes a larger proportion of the elevated blood sugar after IGF-1 falls following from carbohydrate intake. As insulin increases, there are grave additional consequences in the stimulation of eating behavior and the fat making machinery.

Unnecessary carbohydrate intake stimulates eating. This food group's excessive consumption increases the insulin level beyond the nutritional needs of the body. High insulin levels and high carbohydrate consumption create a vicious cycle between the insulin-induced obsession for food and the consumption of carbohydrates. As insulin levels rise, weight gain occurs. **Insulin is the most powerful of all growth factors for fat.** The failure to counsel trapped owners about this fact leads to unnecessary suffering in the western world today.

Carbohydrate, whether it is complex (bread, potatoes, pasta, or grains) or simple (fruit, sugar, corn syrup, honey) break down into one or more of the simple sugars [glucose, fructose or galactose]. Intestinal absorption of sugar is independent of insulin level.

Insulin levels rise after the intestines absorb sugar and dump it into the blood stream. The amount of insulin needed becomes proportional to the sugar load absorbed and the severity of the IGF-1 deficit. Healthy owners have many times the IGF-1 compared to insulin within their blood streams and need much less insulin to uptake sugar fuel into the body cells outside the liver and fat. Conversely, unhealthy owners need extra insulin because their IGF-1 levels have fallen (one cause of insulin resistance). Whatever the insulin secretion rate the liver always receives the highest insulin message content because of its relationship to the portal vein and pancreatic release. The extra insulin increases the fat maker message within the liver and fat cells. The other food groups, protein and fat, need very little insulin for cellular uptake of these fuel types. When enough insulin releases into the portal vein, to bring the total fuel nozzle hormones up to a sufficient level, the blood sugar returns to normal levels.

The greater the IGF-1 deficit, the more insulin needed to process a given amount of sugar intake.

The insulin resistance problem compounds when carbohydrate consumption takes place without potassium. Processed food diets contain less potassium relative to carbohydrate consumption. Real food diets (natural and un-adulterated foods) will contain high potassium relative to carbohydrate content (see mineral table). When owners pay attention to this, tolerance for carbohydrates increases. Sufficient potassium needs to be present to lower the insulin requirement for the amount of carbohydrates consumed. Potassium proves crucial to take sugar out of the blood stream. Only real food contains sufficient potassium with carbohydrate. Processing food removes potassium and adds sodium. Processed food contains reversed mineral proportions compared to body design. Altered minerals within some middle-aged owners bodies causes **low voltage cell syndrome**. Low voltage cell syndrome manifests as weakness and fatigue. Until these owners are counseled on how to correct their mineral imbalances they will continue to grow old at a rapid rate.

Ignoring this simple fact about potassium-depleted foods prevents understanding another cause of **insulin resistance**. Insulin resistance means that more insulin needs to be secreted for the carbohydrate load ingested to return the blood sugar to normal. The blood sugar returns to normal with less insulin when there is sufficient potassium available. One potassium molecule with the help of insulin or IGF-1 moves one sugar molecule out of the blood stream.

Potassium deficiencies create a big component of health issues that begin in middle age. The degenerative processes that begin with potassium deficiencies in middle age often contribute to obesity, abnormal cholesterol profile, decreased

steroid tone, high blood pressure, nervous irritability, diabetes, and decreased energy levels. Middle age disease can sometimes be avoided when the digestive relationship between potassium intake and carbohydrate intake is realized.

A Few More Words on How to Dismantle a Protein

As was mentioned above, the various amounts and sequences (the chain) of 20 different amino acids make up the different protein types. Nutritional value ceases, for building block purposes, when the digestive tract fails to precisely dismantle protein molecules. Precision dismantling proves as a prerequisite for obtaining the continuous supply of amino acids needed for ongoing rejuvenation projects.

Additional concerns arise if consumed proteins partially digest (fragments that contain three or more amino acids). The absorption of amino acid sequences three or more long has potential to activate the immune system. This fact forms the basis for how food allergies start in infants fed foreign proteins before their digestive tracts develop a sufficient barrier. When an undeveloped digestive tract cannot digest protein completely, it will leak the partially digested protein fragments into the blood stream. Once these protein fragments float in the blood stream they will activate the immune system of an infant. When the immune system activates an allergic response becomes possible. Activation occurs when it 'sees' the same sequence again from the ingestion of the same food. The immune response occurs and manifest as food allergy.

Partial digestion and absorption of protein fragments can be a powerful immune system activator against the body, autoimmunity. Autoimmunity means that the body's immune system attacks its own body structure. One cause of autoimmune disease occurs when certain processes injure the integrity of the tube of the digestive system. When the tube integrity compromises it will leak undigested fragments that normally do not penetrate until they are completely broken down. Examples of these diseases are rheumatoid arthritis, systemic lupus erythromatosus, and some thyroid diseases. Some clinicians call this **'leaky gut syndrome.'**

Defects in the inner protective layer of the digestive tube facilitate leaky gut syndrome. Glycosaminoglycans (GAG) compose this protective layer. As this layer continually forms in the digestive tube and degrades, it creates mucus overlying the new GAG layer. Some scientists call the GAG layer the matrix. In fact, through out the different linings of the body such as the skin, the respiratory tract, the inner lining of the blood vessels, the GAG layer forms a protective barrier in these areas. In addition, it is this architectural support meshwork (GAG) that inflates cells, forms cartilage and connects cells together. All of these are formed by GAG. Paradoxically, mainstream medicine often fails to discuss or consider this important layer in the maintenance of health. This is folly because as GAG diminishes so does the owners health.

In order for doctors to appreciate all these facts about GAG, they need to know all the many alias names that it goes by within the literature. Some of them are: cell coat, mucopolysaccharides, basal lamina, ground substance, cytoskeleton, matrix and cartilage.

Since GAG formation is so important to one's overall health, pertinent connecting facts for what determines its formation rate are in order. The amount of IGF-1 determines a particular body local's amount of GAG formation rate. Many physicians reman unaware of this fact because the vocabulary uniting this association has been changed. The older literature described IGF-1 as sulfation factor.

Sulfation describes the critical step in the formation of GAG building blocks called either galactosamine or glucosamine where sulfate adds to them. As long as sufficient IGF-1 instructs the locals listed above then adequate GAG forms. GAG cannot form properly without sufficient sulfation on each glucosamine and galactosamine. The addition of sulfate powerfully sucks in water. The more adequate the water content, in a body area, the less shrinkage. It is largely the GAG content of the body that opposes the forces of gravity, which desires to squish cells flat. Only when there is a sufficient presence of the highly charged

GAG's will there be the ability to impede shrinkage imposed by gravity. Shrinkage occurs from drying out and has been recognized since antiquity to be fundamental to the aging process. Remember, IGF-1 levels depend on adequate growth hormone release, a healthy liver and adequate androgen telling the liver to make IGF-1. If any of the above three requirements fail health suffers.

Physician's Sidebar

A similar vocabulary change preceded the old name of IGF-1 being called sulfation factor. This name change helps to propagate yet another disconnect in physicians minds regarding IGF-1's important role in sugar metabolism. This older name calls IGF-1, the nonsuppressible insulin like activity of the blood. **The old name sheds light on IGF-1's important role in sugar metabolism.** The new name helps to instill fear of the same molecule being a growth factor for cancer cells. Never mind the fact that all healthy people have high levels of IGF-1. Change the name and you can better manipulate how physicians think.

Another powerful disconnect results from the fact that insulin amounts, in almost all medical textbooks, are measured in microunits. However, IGF-1 is measured in nanograms. This helps to perpetuate the concealment of the fact that IGF-1 occurs at levels many times greater than insulin. If more physicians were aware of this fact they would begin asking why the body design would do this? The next step would be economic disaster for the medical industrial complex.

Normally, the digestion process disassembles larger molecules that would activate the immune system. The immune system activates against foreign sequences of protein. Viruses, bacteria, fungus, and partially digested proteins in food contain foreign protein. The body circumvents protein meals containing foreign protein amino acid sequences by dismantling these proteins into their constituent amino acids. Individual amino acids are too small to activate the immune system. These small building block molecular parts, (amino acids) transport in the blood stream to different cells. Once inside the cells, the individual amino acids arrange into new proteins. In this way, amino acid molecular parts organize to create complex structures (proteins) in the cells without immune systems activation in the blood stream.

A Component of Old Age Results from Old Proteins

Old age or chronic disease occurs when the integrity of the individual proteins begin to breakdown. It helps to think of it as the animals that make up the imaginary chain as they become deformed, acquire worn down edges, or attach to debris. The diminished ability to replace proteins as they break down and wear out defines a crucial process of aging. Poor nutritional absorption that results from lack of efficiency in the digestive tract or misguided protein choices cause decline in protein dependent cellular rejuvenation.

The face of someone old and decrepit results, in part, from his or her lost protein integrity. At the molecular level this is nothing more than damaged individual amino acids, which alter the overall effected protein structure and function. The functions of protein encompass their role in cellular support. Support functions are compromised when the underlying protein shape becomes deranged. Deranged proteins have a diminished ability to perform other functions like work and trash removal. These faces, at a molecular level, have accumulated cellular garbage, poor informational substance message content, decreased cellular electrical charge, and accumulated cell rust.

All owners will head this way more quickly without proper nutrition motivation. Better daily nutrition choices that impact cell regeneration and healing provides a place to start. Optimal digestive juices and operative digestive structures encompass integral parts of healing.

Surprises About Where Protein Comes From

Approximately 50% of digested protein comes from the meal content; 25% from proteins within the digestive juices that digest themselves after they perform their roles; and 25% comes from sloughed cells that line the intestinal tube that are also digested. The intestinal tube lining cells have a life span of only 3-5 days. They are sloughed into the tube. Digestive juices dismantle them into their original building blocks. These molecular parts recycle into cell rejuvenation programs.

The individual amino acids comprise the interchangeable molecular building parts of proteins. As long as a building block (amino acid, fatty acid, vitamin or cofactor) remains structurally intact it is free to be used or recycled into any cell synthesis or rejuvenation project again and again.

The ability of an amino acid to interchange terminates when its structure becomes damaged. Replacement of a molecular part only becomes necessary once wear and tear weakens molecular function. In healthy owners, their cells constantly dismantle damaged proteins and efficiently replace them with new proteins.

Healthy bodies contain the sufficient ability to supply the much-needed molecular replacement parts. Adequate replacement parts facilitate efficient cellular trash removal systems and adequate informational substance content. Fat is another molecular building part that the body requires for structural integrity.

Some Important Facts About Fat and a Few More That Mysteriously Have Been Kept a Secret

Four Important Roles for Body Fat (lipids)

1. Cell structure and water retention (simple fats and cholesterol)
2. Hormone precursors
3. Fuel
4. Brain fats keep nerve cells happy

Specialized Lipids (fats), fatty acids, triglycerides, and cholesterol make up the basic structural fats of the body. These building parts are often used as covering for trillions of cell membranes. These occur as part of the cell covering structure (plasma membrane) and organelles (cell factories). These structural fats integrate between varying combinations of interspersed proteins. Both the simple structural fats and the more complex, specialized structural fats compose membranes. The complex group contains lipids with extra molecular parts like phosphate or choline. Ultra-specialized fats have additional properties in body function by their role in specialized cells and organelles.

Beautiful skin demonstrates the importance of structural fat.

It polishes the surface and prevents shriveling by retaining water. Prevention of cellular water loss describes a major function of fat. Inadequate fat content, in type or amount, causes or accelerates water loss. Shriveling and cellular inefficiency occur when water content decreases.

Second, certain lipids (fats) behave as informational substances. These can be thought of as hormonal fats and derive from the essential fatty acids. The body cannot produce essential fatty acids. They must be acquired through dietary sources and absorbed by the digestive tract.

Essential fatty acids transform into message carriers (hormones). Hormonal fats have their own message content contained in their precise shape. Hormonal fats promote information exchange between neighboring cells. Scientists call these types of hormonal messengers the ecosanoids. Availability of essential fatty acids, which the body needs to make certain hormonal fats play a role in the prevention of heart disease, arthritis, and inappropriate immune activation.

Ecosanoids can be further divided into three groups: prostaglandins, leukotrienes, and lipoxins. All of these ecosanoids derive from arachidonic acid, linoleic acid, or linolenic acid-the only known essential fatty acids. Dietary choices and the higher level hormones circulating in the body determine what type of essential fatty acid eventually develop into hormones in cells (See *Entering the Zone* by Barry Sears). Consume fats that lead to a prevalence in anti-inflammatory, heart disease preventing, and immune system optimizing types of message content. Sources for these types of essential fatty acids are found in cold-water fish, borage oil, evening primrose oil, algae and vegetables.

In order to have good message content, there needs to be the right higher hormones. Only when there is more glucagon and less insulin hormones will there be the proper direction for the good precursor hormonal fats to be manufactured from the dietary derived essential fatty acids (see *Entering the Zone* by Barry Sears or the hierarchy of hormones in appendix A).

Some owners consume diets that lead to the promotion of heart disease, inappropriate activation of the immune system, and inflammation producing fatty acids (informational substance precursors). These disease exacerbating hormonal fat precursors occur on cell surfaces when insulin predominates. The partially hydrogenated fats (trans fats) also contribute to this process. Some body processes will activate inappropriate message content when the disease promoting types of fatty acids make up an owner's cell membranes. They promote the above diseases because they contain inappropriate message content.

The third role of fat in the body is to serve as fuel. Per gram of weight, fat provides triple the energy content of protein, and double the energy content of carbohydrate. Only optimal hormone content (message content) permits accessibility of this excellent fuel source to the cells for caloric needs. Which hormone dominates the message content determines the fat access availability for fuel needs. If the wrong hormones dominate message content then the fat cell content becomes off limits as fuel. This condition, and all the complications to other body systems, traps the body in obesity (see glandular chapter).

The fourth role of fat concerns the electrical circuitry of the central nervous system. These electrical insulator types of fats possess unique types of components that provide for the maintenance of proper nerve conduction and health. **The fact that the human brain has more fat weight than nerve cell weight illustrates this point.**

Plant Leaves Provide the First Clue that Lipids are Cool

The ordinary plant leaf demonstrates the water conservation role of fat in the cells. The leaf endures hours of hot sunshine without crinkling or wilting under normal conditions. Leaves possess the right amount of molecular surface structure to retard moisture loss. The leaves are rich in lipid-derived wax. Wax is just slightly modified fat. The fat that lines the surface of the body's cells provides similar water conservation properties in the skin cells. Oils float on the surface of water and provide a vapor barrier against water loss.

The lipid lining of the cells serves to keep cells from shriveling up. A shriveled group of skin cells manifests, to the naked eye as wrinkles. Inadequate cell membrane lipid, in types and amounts, provides one mechanism that leads to wrinkled cells.

The faces of owners who adhere to a low fat diet acquire deep creases between the inferior lateral margins of the nose extending downward towards the edge of the mouth. Later, more wrinkles crease these faces. Inadequate fat content accelerates water loss that causes these changes.

The Successful Dietary Fat is Bad Misinformation Campaign

Lipid (fat) has been given a bad name by the medical industrial establishment. For many years, people have fallen victim to disjointed information. This happens when the dogma for sale and literature dispensed by the sanctioned scientific community is sensationalized.

Availability of information has a profound influence on the practice patterns of medical practitioners today. Some have encountered the medical opinions of men like Bernstein, Atkins, Schwarzbein, and others. Their unique viewpoints led this author to a curiosity about what scientific knowledge had been ignored. Scientific knowledge is often ignored in a profit driven health care system because it is less profitable to make it available for review by physicians and owners.

There is a big difference between diet derived fat versus liver synthesized fat. Owners who acquire fat from too much diet derived healthy types of fat have a lower health risk than owners who gain fat from liver manufactured sources. Fifty years ago, medical research documented that liver manufactured fat was hard compared to many dietary sources of fat.

The liver will only manufacture fat if it receives the informational message to do so. The hormone that directs the liver to make carbohydrate into fat is insulin. Remember, insulin is one of only two of the body's fuel nozzle hormones. When the 'nozzle' correctly inserts, the food groups can be taken up out of the blood stream and into the cellular fuel tanks. When all the cellular fuel tanks are full (this occurs in the well-fed and sedentary state) insulin directs the liver to manufacture fat from diet-derived carbohydrates. Insulin directs the liver to manufacture carbohydrates into fat only when it spills backwards into the blood stream. Carbohydrates spills backward into the blood stream when the cell fuel tanks become full or not enough IGF-1 occurs in the blood stream to help insulin. IGF-1 is the other fuel nozzle hormone. The higher the carbohydrate consumption, the more insulin needed to process the sugar load when IGF-1 levels begin to fall around middle age. The higher the insulin level, the more message content directed at the liver to make carbohydrates into fat (cholesterol and triglycerides).

The higher insulin content within the blood stream, then the higher the 'volume' directed at the appetite center of the brain. The volume control of the appetite center responds to the amount of insulin message content. Insulin stimulates the next feeding urge.

The higher the insulin content, the more "locks" that will be placed on access to stored body fat for energy mobilization. The body is smart and consistent. The overall message content of insulin within the body concerns itself with the storage of fuels. The body can store fuel as fat and to a limited extent as glycogen. An owner's upper limit of insulin production in their pancreas limits the amount of body fat possible. The greater the pancreas production rate of insulin, the more weight gain tendency when dietary indiscretion occurs.

Good Reasons to Fear the Insulin-Directed Manufacture of Fat

When insulin directs the liver to manufacture fat, it is called LDL cholesterol (the bad cholesterol). This type of cholesterol does damage by accumulating in the blood vessels. LDL cholesterol accumulates in the macrophages that line the arteries. Higher manufacture rates of LDL cholesterol, increase the likelihood that this type of fat and cholesterol will accumulate within the arteries. This becomes a likely process in those owners who tend toward high blood insulin levels. These owners eat high carbohydrate diets, possess mineral imbalances or are stressed. The maintenance of health requires that this fact be addressed.

Macrophage collection of LDL cholesterol problems can show up in laboratory test as increased LDL cholesterol and/or as an elevated triglyceride level. For clarity this text will mostly refer to LDL and the reader needs to remember about the triglycerides. In the next few subsections it will be explained why the macrophage cells, being stuffed full of LDL cholesterol, create a problem for blood vessels.

Macrophages Stuff Themselves on LDL Cholesterol

Some owners have a big drain off of LDL cholesterol out of the blood stream and into their fat cells. Their blood fats (LDL cholesterol) stay down even when they eat insulin-producing foods. Not all owners are insulin sensitive. Owners who can eat carbohydrates without retaining fat in the belly area (spare tire of middle age) demonstrate this. Usually the explanation concerns the fact that these owners have a higher IGF-1 level, which means one needs less insulin to process a sugar load. Still other owners indulge in high carbohydrate diets, but have a genetically superior sized 'drain ' (fat cells) once LDL cholesterol dumps from their liver into the blood stream. These owners will become fat, but their abdominal fat cells are so proficient at the removal of liver manufactured fat that the serum levels of LDL cholesterol stay normal.

Most owners are insulin sensitive and will follow the increased insulin message into two health problem areas. For most owners health problems occur in the form of progressive obesity and additional elevation of LDL cholesterol and triglycerides in the blood stream. An elevation of the LDL cholesterol and triglyceride collects in the macrophages that line the blood vessels. These owners do not have a sufficient sized 'drain' off into their fat cells.

The 'drain' describes the exit rate of LDL out of the blood stream and into the fat cells compared to the rate of liver output of LDL production. Deficiencies of either or both Vitamin A and thyroid hormone will cause an additional exaggeration of the slow down in the removal of blood fat.

In a certain proportion of these insulin sensitive owners the chronic increased carbohydrate load, falling IGF-1 levels, sedentary lifestyle, mineral imbalance and their body size combine to exhaust the ability of their pancreases to make enough insulin. At this point, the blood sugar starts to rise. Scientists call this situation adult onset diabetes (90% of all diabetic victims). Some owner's pancreases manage to manufacture and secrete all the insulin needed to keep their blood sugars normal. Both of these groups go on to develop heart disease however and most remain unaware that high insulin levels cause their blood vessel disease problem (see the liver chapter).

Why the Body Needs Cholesterol Supplied Through the Diet

Cholesterol is just a type of fat. Each type of cell requires a certain amount of cholesterol for structural integrity. Cholesterol content in a cell is meticulously regulated to prevent destructive cellular health consequences. Cholesterol provides the building block molecule from which all steroids are made from. The same enzyme that makes cholesterol, HMG Co A reductase, also makes Co enzyme Q 10 and dolichols (the molecule that holds muscle cells together when tension occurs). The popular statin drugs therefore poison all three processes.

Liver manufactured fat is always packaged with cholesterol and specific proteins before being dumped into the blood stream. Once the cells have absorbed their maximum load of cholesterol,

171

there becomes little metabolic opportunity for liver secreted cholesterol. The amount of cholesterol secreted as LDL by the liver increases with rising insulin levels. The metabolically slow fat cells then become responsible for removing LDL cholesterol piece by piece with the help of lipoprotein lipase. Lipoprotein lipase needs heparin as a cofactor for activation. Mast cells secrete Heparin.

The Big Difference Between Liver Manufactured Fat and Dietary Fat is a Secret

Fat is always packaged with varying amounts of cholesterol, but technically cholesterol is a modified fat molecule. Fat and cholesterol need to associate with protein in order to float in the blood stream. The variable that makes a fat available to the metabolically hungry cells concerns the construction of carrier proteins.

There is a distinction in the difference between carrier proteins occurring between the liver manufactured fat and fat from the diet. Diet derived fat has less potential to accumulate in blood vessel walls. The carrier protein of dietary fat contains a protein package that has less potential to plug up a vessel. Where fat originates provides the crucial difference between transport packages. When fat-cholesterol-protein complexes originate in the digestive tract following a meal, the metabolically hungry cells easily absorb them. Remember, insulin is the fuel storage hormone. When insulin directs the liver to make cholesterol-fat-protein complexes, they are designed for storage and not immediate fuel needs. One more way the body proves smart and consistent.

The insulin message content concerns itself with fuel storage in the body. The majority of fuel storage occurs as fat. Normal amounts of diet derived fat form complexes that are rapidly assimilated by the metabolically hungry cells (skeletal muscle, cardiac muscle, kidney, etc.). In contrast, liver manufactured fat is designed to form complexes that are not readily cleared from the blood stream except by fat and macrophage cells. Insulin directed fat-cholesterol-protein complexes are less able to serve as fuel. They are designed at the direction of insulin, the fuel storage hormone. Again, the body is smart and consistent. The consistent message of insulin causes greater amounts of stored fuel (glycogen and fat). The insulin directed fat-cholesterol-protein complexes are designed with fuel storage in mind.

The Unique Way Fat Absorbs from the Diet Illuminates the fallacy of the 'Fat is Bad Campaign'

A physiologic fact clarifies the lack of stickiness of diet-derived fat on vessel walls. Dietary fat and cholesterol absorb via the lymphatics, which differs from all other nutrients. This means that lymph vessels receive the highest vessel concentration of dietary cholesterol and fat. Atherosclerosis (the growth of fat within vessels) of lymphatic vessels is unheard of. Hemodynamically (the flow of fluid within a vessel), lymph fluid moves in a sluggish manner compared to blood flow within the arteries. If there were anything sticky about diet derived fat, it would be more pronounced in the sluggish lymph vessels.

The difference between the type of cholesterol-fat package that blood vessel walls are exposed to compared to the cholesterol-fat-package that lymphatic vessels transport following a fatty meal explains this fact. This difference explains the true cause for the majority of heart disease: excess insulin. The insulin directed, liver manufactured cholesterol-fat packages collect in blood vessel walls within the macrophages. Anatomically, the lymphatic vessels are not exposed to insulin directed and liver manufactured cholesterol-fat packages. They are only exposed to diet derived cholesterol-fat packages that do not deposit along lymph vessel walls.

If only medical schools taught that **the body is smart and consistent**. Insulin directed fat and cholesterol particles are constructed with storage in mind. Long-term storage sites for these types of particles are found in the liver, blood vessel macrophages, and fat cells. Conversely, diet derived fat and cholesterol particles, the chylomicrons have a design that readily allows removal by the body cells and they are stored short term or used as fuel.

Even though lymph vessels are sluggish in flow characteristics and exposed to very high levels of diet-derived fat and cholesterol, they do not plug up. Anatomically the small intestine delivers fat and cholesterol into the lymph vessels following a meal. The protective factor for lymphatic vessels involves the fact that they have no exposure to liver manufactured cholesterol-fat-protein complexes (LDL). Intestinal lymphatic vessels eventually drain into the left neck area, the thoracic duct. In this location the fatty parts of digestion dump into the blood stream.

In contrast, other products of digestion such as protein, carbohydrates, minerals, and vitamins, etc., dump directly from the intestinal cells into the portal vein. The portal vein collects all blood and nutrition (except fat and cholesterol) from the digestive tract and takes it straight to the liver.

The blood vessels clog up whenever insulin levels rise to the point of directing the liver in excessive manufacture of the sticky type of cholesterol-fat packages or LDL cholesterol. This is sticky because it is taken up by the macrophages that line blood vessels. Macrophages are immune system scavenger cells that under well-fed circumstances bed down on blood vessel linings and stuff themselves with LDL cholesterol.

When the liver creates excessive fat-protein-cholesterol complexes (LDL cholesterol), they incorporate into macrophages in abnormal amounts. Excessively stuffed macrophages with LDL cholesterol grow and grow. Year after year they continue to accumulate LDL cholesterol and eventually form giant cells called foam cells. Foam cells are the earliest lesions recognized by scientist as the beginning of heart disease.

Some of the places that macrophages bed down and become foam cells occur in the coronary arteries, hearing apparatus, kidney vessels, peripheral leg vessels, and retina. These tissues prefer fat for fuel. The LDL cholesterol accumulation trouble compounds with a sedentary owner.

High insulin levels in sedentary owners allow LDL cholesterol to collect in the macrophages continuously. Theses cells permit fat to collect due to lack of exercise. After many years, this vicious cycle allows fatty streaks to grow enough to block blood flow. Blood flow can suddenly worsen in these narrowed areas when other factors promote blood clots to form. When severe enough, the constriction causes cell death to cells that are beyond the blockage. The other common mechanism for the sudden loss of blood flow is vessel spasms in the narrow area. These two climaxes are the typical scenario for the number one killer of Americans past middle age, heart disease. The earlier available information on this subject has been disjointed and fragmented.

High insulin producing bodies manufacture more fat and cholesterol. Insulin directed cholesterol-fat packages are designed for storage depots. These packages designed for the storage depots of the body are slow to drain off out of the blood stream. These individuals have an increased risk for this type of heart disease when they are living sedentary lifestyles, have unmanaged stress (see stress chapter), contain mineral imbalances, suffer a fall in IGF-1 levels and/or consume high insulin promoting diets. All five of these risk factors increase insulin levels. Increased risk arises from the increased tendency for macrophages to ingest more LDL cholesterol than is healthy.

The relationship between high insulin levels because of falling IGF-1 levels, high stress levels, reversed minerals ratios diet, high carbohydrate intake and sedentary lifestyles are the crux of the problem for the majority of all heart disease victims. The other risk factors for heart disease development accelerate this primary disease producing mechanism in the blood vessels. Rather than focus on the hopeless mantra about some owners genetics, empowerment comes from a focus on how dietary and lifestyle changes prevent the problem.

Some owners have a greater tendency for heart disease based on primitive survival strategies. A survival advantage was obtained in prehistoric times when a body could store fat in times of plenty in order to survive periods of famine. Owners that made more insulin had a survival advantage. The food supply was never predictable so there was no accumulation of blood vessel fat.

The modern day food supply continues uninterrupted in supply to these same survival equipped insulin producing machines. Add in the greatly increased insulin requirement that becomes necessary to handle a chronic processed food diet. This occurs secondary to its diminished potassium content. These same survival equipped owners are now fat producing machines. **An owner still has the ability to turn off his/her fat making machine. The liver fat making machine turns off by consistently choosing a low insulin requiring diet, obtaining ample potassium, minimizing sodium, doing aerobic exercise, correcting the other fat promoting hormones, and stress management (as suggested in earlier chapters). Instead of helping owners understand this, they are led down the erroneous path of what is for sale by the complex.**

Unfortunately for the average owner little incentive exists to share (publish and sensationalize) this basic scientific understanding. The 'fat is bad campaign' contains largely hype that allows the relationship between insulin and blood vessel disease to remain a secret. Knowledge about the insulin secret jeopardizes the more lucrative approaches for treating blood vessel disease. However, the lucrative approaches contain side effects and toxicities that if commonly known would create even more of a reluctance for these symptom control methods. Symptom control, so prevalent within the mainstream paradigm, has nothing to do with healing. Passive owners are stuck with symptom control and all its unfavorable side effects. However, what about those owners that possess the focus to commit to a path of healing?

The Weak Spot in the Complex's Campaign Against Dietary Fat

The 'fat is bad' media campaign has a weak spot. Simply monitoring the cholesterol profile while on a high good fat/protein and low carbohydrate diet exposes the weak spot. This dietary approach directly contrasts with the official campaign against fat and in favor of a high carbohydrate diet.

The owners, who try a high protein/fat and low carbohydrate diet for three months or more while they defer judgment, will likely see several improved health benefits. At the end of this time period, each owner should assess the improved mental clarity that occurs because less 'brain fog' occurs. Hypoglycemia induced by high carbohydrate diets has stopped. They will also notice a decreased appetite due to the fall of insulin levels. There is less brain appetite center stimulation for the next feeding event. There will be consistent weight loss because of lowered insulin levels that frees the 'locks' on the body's access to fat stores. Finally, there is a dramatic improvement in the cholesterol profile due to less insulin message content instructing the liver to make carbohydrates into LDL cholesterol.

Attention to thyroid function, adequate androgen output from the adrenals and gonads (testis or ovary), optimal cortisol output, and IGF levels are important for normal fat parameters. Aerobic exercise should be optimized enough for the muscles, skeleton, and cardiovascular system to be utilized to full capacity. These changes transform an owner onto the longevity track of life.

The high protein/good fat and low carbohydrate diet becomes less controversial when reviewing the epidemiological studies that have noted the absence of heart disease in cultures that consume a low carbohydrate, but high fat/protein diet. Eskimos exemplify this fact. Eskimos also exemplify the relative safety of obesity acquired through over indulgence in fat intake. Many Eskimos are obese because of the arctic climate. The type of fat Eskimos make is opposite to liver manufactured fat that arises from a high carbohydrate diet. When these same Eskimo peoples begin to adhere to a high carbohydrate diet, after about 20 years they develop heart disease at rates similar to their western contemporaries (Atkins 1999).

This correlates with the absence of heart disease in America until the 1900's. Up until this time, the general population survived on a high fat and protein diet (chicken, meat, eggs, and beans). The carbohydrates that were available were unprocessed and hence contained a high potassium content. Remember, when the potassium content is high, the need for insulin diminishes. After the turn of the century, as processed sugar and cereals became available, the 20-year rule began to tick. This epidemiological tool measures the time between a change in cultural behavior and a public health effect. By the 1920's heart disease had become an epidemic in America. (*Atkins, 1999*)

On the Other Hand: Chemically Reactive Fats in the Diet Cause Heart Disease

The consumption of unhealthy fats are not allowed in high fat diets if one wishes to be free of blood vessel disease. The people of the world free of heart disease, while they indulge in a high fat and low carbohydrate diets, are not the ones that consume high amounts of chemically reactive fats. Chemically reactive fats have the scientific name, partially hydrogenated vegetable fats. This distinction about good fat and bad fat is often overlooked and leads to confusion in the minds of clinicians as well as owners in regard to the safety of high fat diets. Diets made up of real fats are safe while diets made up of large quantities of chemically reactive fats are not. Chemically reactive fats contribute to oxidation (rust) in the blood vessels.

The chemically reactive fat groups are chemically altered polyunsaturated fats. The chemical process of hydrogenation has altered them. The hydrogenation process causes a twisting deformation that is an unnatural occurrence in real fats. Numerous baked goods, margarines, breads, snacks, and chips have altered fats added to them. The hydrogenation process provides one example where altered fats add firmness to the product. These polyunsaturated fats are chemically altered and become more twisted when they stack together. Fats in the body need to be stacked together in the cellular membranes (the lipid bi-layer). The twisted configurations of chemically altered fat means they form awkward fat conglomerates (irregularities) that line the cell membrane.

Endothelial cells fit together like tiles on the interior blood vessel lining. These cells, like all cells, have a cell membrane made up of fat interspersed with proteins. The polyunsaturated partially hydrogenated fats are also chemically reactive with many oxidizing agents that may be in the blood stream (aluminum, fluoride, oxygen radicals, carbon monoxide, ozone and other smog components, etc.). When these fats oxidize, they promote 'Velcro' formations on the inside wall of the blood vessel. This formation summons the macrophages to lay down a temporary patch job that sometimes is never repaired. When LDL cholesterol is high, these macrophages collect fat and eventually grow into foam cells.

A component of blood vessel health depends on whether the inner lining layer (the tiles) of the blood vessel wall composes itself with the chemically altered polyunsaturated fats (hydrogenated fats) or healthy fats. The diet determines this content. Diets consisting of fast food fats, margarine, fried foods, baked goods, and commercial breads will add twisted fats to the lining layers of that owner's blood vessels. Diets that contain real fats promote surfaces in the blood vessel, which are composed of the less chemically reactive fats (real fat).

Picture two beautiful flowerbeds. In one, short stocky stems grow and in the other fragile, slender stems grow. The contrast between the two types of stems allows an analogy that elucidates the body fat composition type problem.

Short stocky stems are analogous to blood vessel friendly fats (olive oils, fish oils and canola oils). They are more durable. Good fats lining one's arteries only become possible when the owner makes appropriate diet choices when consuming fats. Fragile long stems are analogous to chemically altered polyunsaturated fats (found in junk food, processed food, and fast food). They are more vulnerable to numerous oxidizing agents that occur in the blood stream.

Visualize a dog (oxidizing agent) that sequentially runs around in each type of flowerbed. The short stocky stems would sustain less damage because of the differences in structural characteristics between the two types of flower beds. The oxidative vulnerability in the blood vessels is analogous to what happens in the blood vessel lining cells when unhealthy fat is eaten. The vulnerability of blood vessel lining cells to trampling by the "dogs" being unleashed in the blood stream are determined by the composition of the "stems." In a dirty environment where air pollution, water pollution, and food contaminants prevail, the blood vessel lining cells are overexposed to oxidizing agents (rust producers). The rust producers are analogous to the dog running around in the flowerbed. Real fats versus altered fats are analogous to the durability of the stems in the flowerbed. The more chemically reactive (altered) the fats become, the more they create rust (Velcro formations). The more the lining cells in the blood vessels compose themselves with real fats, the more durable they become when confronting oxidizing agents.

This consideration gradates the relative risk of different types of dietary fat and their potential to produce disease. The amount of abdominal fat serves as a clinical marker for diet and/or lifestyle that leads to higher insulin levels. Men who have a greater abdominal measurement than hip measurement evidence high insulin levels. Women who are greater than 80% their waist compared to hip measurement evidence high insulin as well. Until insulin levels become optimal, by diet and lifestyle changes, health deteriorates.

The higher the oxidant exposure load that operates on a daily basis, the more important it becomes to minimize consumption of the chemically reactive fats. Owners who are chronically exposed to air pollution, water impurities (fluoridation), and impurities in their food will have an increased rate at which their chemically reactive fats oxidize. Chemically reactive fats oxidize when exposed to harmful pollutants. Improving dietary fat choices and avoiding environmental 'rust' producers benefit healing.

Liver Manufactured Fat Clogs Blood Vessels

It has been known for years that LDL cholesterol synthesis rates are controlled by the balance between the levels of insulin and glucagon. The insulin message tells the liver enzyme, HMG CoA reductase, to make sugar into cholesterol. Glucagon's message instructs the liver to stop making cholesterol by inhibition of this same enzyme. Glucagon also stimulates the release of fuels into the blood stream. For this reason, healthy, active people with a higher glucagon level will have lower cholesterol levels.

Glucagon will not help people who are sedentary and obese. Glucagon stimulates the dumping of fuel into the blood stream from the liver. In chronically sedentary types, the extra fuel stimulates the release of insulin, eventually. This effect provides an example of the weight and counterweight system that keeps opposing hormones in balance. People who sit around all day, but eat the right diet initially promote the release of glucagon. However, they lose the benefits of glucagon when insulin needs to be secreted to manage the extra fuel that glucagon directed to be made and secreted into the blood stream.

Extra insulin arises because sedentary owners have little need for increased fuel in the blood stream that glucagon directs. If owners on glucagon promoting diets were to exercise, they would use the extra fuel and need less insulin. Lower insulin relative to glucagon signifies the liver enzyme, HMG CoA reductase, will stay less active and hence less cholesterol is manufactured. When this happens, the serum cholesterol level decreases.

Owners have a choice about taking toxic medication to lower their cholesterol or to heal. Medication often becomes unnecessary if they begin a program that helps the liver to work properly. Practices that raise glucagon and lower insulin keep cholesterol within the healthy range for the vast majority of owners. A high protein/good fat, low carbohydrate diet promotes the optimal ratio between insulin and glucagon. The proper balance of these two hormones is difficult for sedentary owners because of the need for increased insulin (see above).

Additional Hormones That are Beyond the Diet, But Influence Digested Foods to Deposit as Fat

Most owners who take active steps in their daily exercise and dietary decisions will notice a dramatic fall in their blood cholesterol. Their glucagon and insulin ratio have been favorably altered. Adequate message content from other body hormones are also important - androgen (from adrenals and gonads), estrogens, thyroid hormone, IGF-1, and cortisol. When these are correctly proportioned blood fat-cholesterol-protein complexes will benefit.

These considerations can be assessed (except for IGF-1) through a complete 24-hour urine test that includes a quantitative and qualitative analysis of what hormone levels are manufactured. Healthy owners have optimal amounts of critical steroid and thyroid hormones that pass into their urine over a twenty-four hour period. Unhealthy owners pass sub-optimal amounts of critical androgen steroid hormones and possibly excessive amounts of estrogen and stress steroids. If thyroid function is inadequate, the urine is often the most reliable way to detect the deficiency. These hormone considerations become important when lifestyle and dietary modification fail to improve the cholesterol profile.

The need for a twenty-four hour urine collection illustrates what science has revealed versus what is practiced. Hormone levels can fluctuate widely in the blood stream. Some hormones have life spans measured in seconds. Most have life spans measured in minutes. When a blood sample is taken, the amount of hormone measured in that sample reflects that bloods informational content at the instant the venous sample was drawn. This method provides an accurate measurement only if the hormones in question are relatively stable in the blood stream.

Most hormones vary widely throughout the day in the blood stream. The different informational substances convey a precise message to the body cells. Different activities of daily living obligate the release of different hormones that communicate to cells on how to spend their energy. Each of the varied life situations requires different informational direction of the body energy quotient. Therefore the hormone levels and types during stress, pleasurable activity, fasting, or post exercise will predictably be different. Different activities and emotional states require a change in the blood stream information to accommodate the change in body energy direction required.

An accurate method to assess hormone tone and pressure would incorporate the basic scientific understanding that over a typical twenty-four hours an average amount of hormone output will occur. Urine measurements are useful only when the hormone of interest exits the body via the kidney. This is true for thyroid and steroid hormones and includes the different breakdown derivatives of these different hormones. Urine that is collected in a typical day can more accurately measure the types and amounts of hormones present. This sophisticated approach toward an assessment of the hormone status is more appropriate.

Most of the larger hormones composed of amino acid sequences (peptides or proteins) do not pass in the urine. This class of hormones must be measured in the blood stream. Insulin, glucagon, IGF-1, and some of the larger pituitary and hypothalamic hormones provide examples of this class. If a more accurate method for the assessment of hormone status exists, patients have the right to know about this option.

Physician's Sidebar

Many physicians are unaware that they are victims of an education that has removed some basic scientific understanding that conflict with profit. There have been numerous instances where a patient brings in a well-run twenty-four hour urine analysis only to be scoffed at by their conventionally trained, but limited second opinion physician. Owners are advised to have a little fun with these types of practitioners. Ask them to explain all the big words that describe the different breakdown products of certain hormones as a start to this fun adventure. A good physician at this point will admit that he doesn't understand the results and will look into it. Unfortunately, many physicians that are confronted with what they do not understand habitually resort to attempts to discredit approaches that inquire above their education level. When one confronts the later type of practitioner, it may be time to begin thinking about a new doctor.

Fluids and Electrolytes

Fluids and electrolytes describe the water and minerals that make up the various body juices. Digestive juices contain enzymatic machines that are specific to the digestive chamber where food presents. The mineral component of juices and body processes that have an effect on the adequacy of digestive juices will be covered in this chapter.

Good health remains possible when adequate juices secrete and absorb at the proper time during the digestive process. The digestive system requires the appropriate quality and quantity of digestive juices or degenerative processes occur. These degenerative processes rob the body of vitality when the quality of mineral content contained in the digestive juices deranges.

Each adult consumes about two quarts of fluid every day. In addition to this fluid intake, the digestive tube itself secretes an additional seven quarts into the meal contents. The digestive tube needs to secrete various digestive juices in order to assimilate the various molecular parts contained in one day's food intake. The total seven additional quarts of digestive juices secreted each day needs to be in properly proportioned amounts. These different digestive juices secrete into the tube as various chemical combinations designed to dismantle different food types. In the end, the bowel movement contains less than 1 cup of water. A typical digestive tube encounters two and a half gallons of fluid every day. When the digestive tube remains healthy it reabsorbs not only most of this fluid, but also usable molecular parts.

The Amount of Digestive Juices Secreted in the Tube

Daily ingested fluid: >2 quarts

Daily secreted juices into the digestive tube from the various digestive-lining cells: >7 quarts

Salivary glands >1.5 quarts

Stomach secretions >2.5 quarts

Gallbladder secretions >.5 quarts

Pancreas secretions >1.5 quarts

Intestinal secretions >1.5 quarts

Total fluid load daily passed into the digestive tube is >9 quarts

Total reabsorbed fluids daily farther down the tube on average is 8.8 quarts or diarrhea manifests itself.

Jejunum 5.5 quarts reabsorbed
Ileum 2.0 quarts reabsorbed
Colon 1.3 quarts reabsorbed

The mineral content of the digestive juices and meal form a determinant for the ability of the body to reabsorb the fluid in the digestive tube before it arrives at the rectum. Diarrhea results from too much fluid. Constipation results from too little fluid reaching the rectum. Fiber and certain minerals keep water in the tube and promote a loose stool.

Certain minerals or substances pull water with them as they move down their concentration gradient. Scientist call this process osmosis. Osmosis describes a way for predicting which direction water will move. Substances that pull water with them as they move into different body chambers are called osmotically active. These substances actively involve the digestive tube, intestinal cells, kidney filtration system, etc. When these substances move from one body chamber to the next they to pull water with them.

When healthy, the act of absorbing the molecular building parts contained in a meal pulls water back out of the digestive tube. Bowel movements of healthy owners do not contain excessive osmotically active material and therefore, their stool contains little water. Deficiency or excess in osmotically active bowel contents leads to constipation and diarrhea respectively.

Indigestible fiber has a weak osmotic effect that allows sufficient water to stay in the bowel movement and facilitate regularity. Sources of indigestible fiber are vegetables, fruits, and psyllium husks. Some minerals incompletely or poorly absorb so they osmotically retain water. Salts of poorly absorbed and osmotically active minerals or substances are the salts of magnesium, vitamin c, and some antacids.

Part of the task of efficient digestion involves adequate mineral absorption. Minerals usually present to the digestive tract in the form of a salt. A salt signifies that positive and negatively charged minerals occur together. The positively charged mineral of the salt is more important and always named first. Examples of commonly ingested salts are ferrous sulfate, calcium citrate, magnesium oxide, potassium chloride, and sodium chloride. When a salt absorbs into the body from the digestive tract it will pull water with it. When a salt fails to absorb in the digestive tract it will retain water within the tube. This fact explains why some mineral salts have a laxative effect.

Calcium and iron are important minerals. The salt they hook on to (the negative half) will determine how readily they absorb into the body.

Most owners are aware of the importance of obtaining adequate calcium and iron. Often times this opinion results from an effective media campaign. The more important a mineral is, the more potential it has to harm body tissues if the mineral fails to be carefully regulated. Calcium and iron illustrate the importance of mineral control within the body tissues.

Calcium Can Help or Kill Cells

Calcium intake requirements are on the mind of most owners. Calcium plays an important role in the maintenance of adequate bone mass. Many owners ingest some form of calcium salt on a daily basis. There has been very little attention given to taking too much calcium when certain factors exist. Calcium intake can accelerate the aging process.

Calcium is controlled and channeled in the tissues of the body by an elaborate regulatory system of hormones. Calcium content must be narrowly channeled. Therefore, an elaborate system of calcium regulatory hormones becomes necessary. Calcium can be better understood when one realizes that all cells operate as miniature batteries. Maintaining a concentration difference between calcium and magnesium charges the cells. Calcium is pumped outside the cell while magnesium is pumped inside the cell. Calcium occurs at high concentrations in the fluid around the cell because of this pumping around the membrane (a ratio of 12,000:1). Magnesium maintains a relatively high concentration inside the cell.

Adequate cellular energetics (adequate ATP formation) maintains this electrical gradient between these two opposing minerals. A similar gradient occurs between sodium outside the cell and potassium inside the cell. Cells constantly drain this gradient between these opposing minerals. The energy released is used for cell work. Energy released by the draining process is like any activated battery. In the case of cells, the energetic gradient is constantly drawn upon and used to live. Simultaneously, the healthy cell constantly recharges the gradient difference by utilizing ATP derived from the cell power plants (mitochondria). Sufficient ATP recharges the membrane gradient between these opposing minerals. Specific mineral pumps maintain the concentrations differences of these minerals around the membrane. Mineral pumps are located in the membrane and require ATP to power them. The higher the concentration differences between the opposing minerals, the higher the amount of energy available for the cells.

The differences in concentration between calcium and magnesium about a cell's membrane allow the performance of useful cellular work. The concentration gradient also prevents harmful ions from penetrating the cell and causing damage within. Healthy cells are able to generate large concentration gradients between calcium and magnesium and also between sodium and potassium. They achieve this concentration gradient by having highly functional mineral ion pumps in their membrane. These pumps exchange opposing minerals against their concentration gradient. **This battery charging process occurs in all the body's 100 trillion cells.**

The large concentration difference between opposing minerals around the membrane creates electrical potential. The energy contained in the membrane drains down to perform cellular work. The same principle applies to any battery when it becomes the mechanism for donating energy to power a toy. Also, both types of batteries (the cell and the toy battery) have to have a way to recharge that prevents depletion.

All batteries, whether living or manufactured, need to keep their mineral content within carefully gated channels. When the channels are circumvented battery corrosion results. Calcium needs to be channeled into and out of cells by carefully regulated channels. The cell performs work when calcium enters through specific channels. The pumping of calcium outside the cell and magnesium inside the cell requires ATP energy. Sufficient ATP allows the cell membrane to recharge as it depletes.

If calcium gets inside the cells inappropriately (outside the gated channels), it harms intracellular structures (molecular corrosion). Once inappropriate calcium gets inside a cell it is difficult to remove it. As owners age, inappropriate calcium sneaks in and chemically reacts with delicate intracellular structures. If enough calcium gets inside a cell, the mitochondria begin to sequester it. Eventually extra calcium causes the mitochondria to swell and weaken. Weakened mitochondria become less capable in their role as a power plant facility (ATP generation). Inappropriate entry of calcium into a cell irreversibly deactivates enzymes.

When calcium enters a cell through the appropriate channels (gate), it does not have a destructive effect on the cell. The movements of calcium, through these channels and down its concentration gradient releases electrical energy that the cells use to perform the work of living. In contrast, processes that accelerate inappropriate calcium accumulation in the cell accelerates aging.

Calcium is elaborately contained by a complex interrelated group of hormones. These different hormones occur at diverse entrance and exit sites for calcium throughout the body.

Situations that overwhelm or disrupt these regulators of channeling calcium safely through and in the tissues can cause cellular harm.

Some of the calcium regulatory hormones and proteins are worth mentioning. Calcitonin hormone opposes parathyroid hormone. Androgen hormones opposes cortisol hormone. Active vitamin D hormone opposes inactive vitamin D. Blood albumin content opposes the freely dissolved blood calcium.

Recall the antique weight scale with a weight and a counter weight. This provides an analogy to a hormone being secreted and the fact that there will always be a counter-regulatory hormone (counter weight). The counter weight hormone always attempts to balance the response. Hormonal imbalance leads to disease due to the breakdown in the balance between the opposing hormones. Loss of balance between the opposing hormones "tips the scale" out of the optimal equilibrium. The loss of hormonal balance within, tips energy usage into unbalanced pathways that lead to wear and tear. Wear and tear manifest in the tissue that receives message content from the out of balance hormonal state.

Body tissue wear and tear all too often results from poor quality message content. Poor quality message content, in turn, results from sub-optimal hormone mixtures. This leads to a disorganized cellular direction on how to spend energy wisely. The balance between the different opposing calcium controlling hormones determines where calcium channels. There needs to be a balance between regulatory and counter-regulatory calcium hormones or tissue injury becomes possible. Inappropriate calcium can enter a cell and injure its delicate intracellular contents.

Certain Minerals Opposition Charges the Cell Battery

The body's system of chemical reactions sustains life. Often times these reactions are powered by the concentration gradient between opposing minerals around the cell membrane. Certain opposing minerals create electrical membrane potential between these minerals. The cell membranes accomplish this difference in mineral concentration by specialized pumps located within its membrane. Each type of membrane pump powers itself by the energy contained within the ATP molecule. ATP availability determines the ability of a cell to recharge its membrane. The better a cell recharges its membrane, the more energy available for cellular work.

One of the main caloric expenditures of the body involves energy used by these various mineral pumps. They are present throughout all cells in the body. Consistent with this fact, large amounts of ATP are necessary to charge the differential between these opposing minerals against their concentration gradient. The more ATP that gets made and used, the more calories burned. Combustion of calories creates ATP within the mitochondria. A major share of the ATP created powers the membrane pumps that create the huge concentration differences between opposing minerals. The concentration buildup of these opposing minerals can be used in the performance of cellular work.

Magnesium in high concentrations in the cell opposes the high concentration of calcium outside the cell. This allows an electrical gradient that the cell can harness to function within that cell type. When a car battery discharges in the wiring of the electrical system of the car it powers many gadgets. The cell "battery" (the cell membrane) powers the activities of life. Constant energy is required to power the cell battery maintained by the gradient between calcium and magnesium (also between sodium and potassium). This is similar to a car battery that needs to be recharged while the auto burns fuel. The combustion of protein, fat, and sugar occurs only after processing into common, combustible derivative called acetate. Acetate is the simplest fatty acid.

When acetate is exposed to oxygen within the power plant [the mitochondria] of the cell, the energy released gets trapped as ATP. ATP can then be used to recharge the cell membrane by powering the mineral pumps. This describes the process that continually occurs throughout life in the trillions of cellular batteries contained within the body. Energy is contained in the membranes and is created by the concentration differences between these four minerals.

Healthy cells avoid unnecessary oxidation from inappropriate escape of un-channeled calcium inside the cell. Potential trouble always lurks inside the body if calcium is allowed to bind to an anion (the negatively charged molecules within any cellular protein) that precipitates (forms a solid). Calcium always forms the positive part of any salt that it forms. Some salts of calcium prefer to stay in solid form and do not dissolve well in body fluids. Calcium salts outside of bone tissue are of the dissolvable variety in a healthy body. Even under the best of circumstances, cells confront numerous potential precipitate formers (in-dissolvable salts). If not constantly flushed from body tissues these will form solid salts of calcium. Common clinical examples of this process are found in kidney stones, osteoarthritis, bone spurs, and calcium deposits in soft tissues and blood vessels.

Excessive dietary intake of phosphates and oxalates can lead to acceleration of these solids deposition in the tissues. An increase in water consumption helps the body clear these substances. Soda pop (diet and regular) contains high amounts of phosphate. There is a price to pay for clearing it through the urine. Each phosphate molecule cleared from the blood stream requires an obligatory loss of one calcium molecule. Soda pop contains phosphate, but lacks calcium. In order to remove excess phosphate, there needs to be calcium removed from somewhere else in the body. If it becomes chronic (as in the case of the habitual soda pop user) the bone calcium leaches out of the bones to allow phosphate removal. This shows up clinically on x-ray as premature loss of bone mineral content. Mineral deficient bones are known as the disease osteoporosis.

Energetic situations that allow inappropriate calcium into the cell are also a concern. This becomes more destructive to the metabolically active cells (nerve, heart, kidney, etc.). Dr. Salpolsky from Stanford University has documented low blood sugar in the hippocampus (an area of the brain involved in learning and memory). Low blood sugar in the brain leads to low energy in these nerve cells. The 'force field' diminishes and a consequent massive in flux of calcium can occur. This vulnerability stems from the fact that the brain cells, under most circumstances, can only burn sugar for energy needs. Most other cells readily utilize protein and fat for fuel. This makes brain tissue much more vulnerable to low blood sugar. Low energy in a nerve cell, caused by low sugar availability in this situation, allows the massive influx of calcium. It is the massive influx of calcium that irreversibly harms intracellular contents. In these cases, calcium harms cells because the low energy content within the cell membrane allows it to penetrate the cell through inappropriate methods.

The vulnerability of cells in low energy states can be visualized by imagining the star ship Enterprise in an attempt to ward off the attack of missiles. The star ship Enterprise does this by powering up its force field. This is similar to what any cell in the body must constantly perform in order to keep out inappropriate ions like calcium. Every cell must power up energetically against the influx of calcium (and other harmful ions) through inappropriate channels. Calcium can be an intracellular missile. That is, when it penetrates the cell outside of the appropriate channels (the membrane pumps), it forms solid complexes with cell components.

Other states of cellular energetic depletion occur in situations where the blood or oxygen supplies compromise (drowning or cardiac arrest) or if vulnerable tissue excite beyond its energetic capacity to maintain adequate cellular membrane energy (seizure). All of these mechanisms injure cells because they allow the depletion of the cell membrane's electrical charge (the force field). When the force field depletes, a massive influx of inappropriate calcium enters the cell. Excess levels of calcium react destructively with intracellular components. (*Sapolosky, 1992*)

Calcium's energetic opponent, magnesium, doesn't need the same strict regulatory control within the body. Magnesium doesn't seem to form solids in the tissues, but forms soluble complexes (those that dissolve in body fluids). This difference in behavior, between calcium's chemical reactive properties contrasted to magnesium's, helps explain why the presence of calcium becomes necessarily layered with hormonal protection systems and backup protection systems. Dietary factors can affect the availability of calcium within the body. Different body minerals, which include calcium, are necessary to power up the trillions of cell batteries. The cell batteries are used for the reactions of life and to defend the cell from outside the cell's hostile ions. Calcium can be hostile when the body fails to properly regulate its presence (see mineral chapter or *The Body Heals*).

Iron: Life Giver or Testicle Taker

A successful weight loss requires functional glands. One of the main glands that offer message content to help burn fat is the testicle. Excessive iron accumulation within the testicle will poison its functional ability. The body contains elaborate hormonal systems and back up systems that are designed to prevent the entry of excessive iron into the body. In addition to the testicles, the heart, pancreas and tongue are all sites of excessive iron accumulation. Scientists call the early stages of this disorder, hemosiderosis and later, the more advanced form, hemochromatosis.

Iron is a lot like calcium. The access of iron to various tissues must be precisely regulated by the digestive tract. Iron must be carefully utilized within the hemoglobin molecules of red blood cells to carry molecular oxygen to the tissues. If the multifaceted protective mechanisms of the body fail to prevent excess iron in certain body cells, tissue destruction occurs. Testicles become particularly vulnerable when iron control systems breakdown. Iron needs to be heavily regulated toward balance in the healthy state.

Men lose about .6mg per day of iron and women lose about twice this amount during their menstrual cycle years. A body will carefully attempt to match iron absorption rates to loss rates in order to maintain iron balance. Phytic acid, phosphates, and oxalates in the diet bind iron in the digestive tube into solid complexes. The excess presence of these types of substances can prevent adequate absorption of iron into the body tissues that need it.

193

The body needs small amounts of trace minerals. These include iron, zinc, manganese, cadmium, selenium, copper, and nickel. All of these trace minerals need adequate stomach acid to be absorbed. The intestinal power pump requires adequate acid to function. The creator designed a power pump in the stomach that exchanges the trace minerals for one acid proton [H+]. **The power pump maintains adequate trace mineral absorption.** Acid deficiency can potentially compromise trace mineral absorption. Adequate stomach acid also facilitates the less absorbable form of iron [Fe+3] to complex with vitamin C. Only when Fe 3+ is chemically reacted upon will it form the more absorbable form [Fe+2]. It is only the Fe 2+ form of iron that carries oxygen when associated with hemoglobin in the red blood cells. Inadequate stomach acid causes vitamin C to be in its inactive form.

Adequate stomach acid allows iron absorption in the stomach-lining cell. Once iron is inside this cell, a protein, ferritin, designed with iron safety considerations carefully sequesters it. The gut lining cells have a lifespan of only 2-5 days. The body must decide whether it needs this iron or not. If the answer is negative, the iron sloughs off (when the cell dies) into the digestive tube where it passes in the next bowel movement. This mechanism provides the first body defense against iron overload that prevents tissue injury.

Physician's Sidebar

The fact that the stress response leads to blood vessel inflammation, when it becomes chronic, involves the fact that certain inflammatory proteins release. C-reactive protein is only one of several of these types of proteins released within the stress response. Collectively these proteins are known as the acute phase reactants. Other acute phase reactants include: fibrinogen, complement, interferon, haptoglobin, ferritin, and ceruloplasmin. The increase of these protein types within the blood stream, in a setting of physical stress, makes sense. The physical stress of combat or running from a large animal requires the increased activity of the immune system, blood clotting and remanufacture of new blood cell components. Each of the above acute phase reactants contributes to this overall scheme for surviving physical trauma. However, in the setting of chronic mental stress when these acute phase reactants also increase there becomes the increased tendency for blood vessel inflammation. Type A personalities provide one example of how some owners find themselves always in survival mode (the stress response).

The above cause and effect relationship of the acute phase reactants and chronic stress develops a more comprehensive picture. An elevated C-reactive protein develops only one small part of the overall picture of blood vessel inflammation process. Other specific acute phase reactants like ferritin and fibrinogen contribute directly to blood vessel inflammation.

Healing type-A personality individuals' blood vessels becomes possible when they understand the bigger connection of the scientific facts. Rather than falling for the complex's tactics of creating hopelessness about the cruel hand that genetics has dealt so take your cholesterol lowering prescription, healing becomes possible. Healing occurs when a type-A personality is provided insight into the importance of increasing positive emotions and the physical activity level while decreasing the stressful behaviors. As chronic stress tendencies resolve, the acute phase reactants decrease.

The relationship to increased acute phase reactants and a tendency to retain excessive iron may better explain one healing effect of chelation therapy. The chelating agent, EDTA, removes iron. Many heart and blood vessel diseased owners were long ago noted to have elevated ferritin levels. Iron excess is emerging as a new risk factor for blood vessel disease. The higher the serum ferritin levels the higher the iron in the body. As the chelation treatments progress their ferritin levels begin to fall. Could it be that one of the benefits of chelation therapy is that it reduces excess iron and therefore allows the blood vessel to begin healing? Tying together the type A personality, high serum ferritin and the possible benefit from chelation begins to help one think outside the box of the standard but fragmented way for discussing heart disease.

The next layer of body protection from iron is found in the transport protein, transferritin that orchestrates delivery into the various iron storage depots. If iron accumulates to high levels, these storage depots become injured first. The body protects itself from additional iron storage damage by increasing the transferred level in the blood stream. Transferritin levels are sometimes used to measure the risk of this storage problem. Common storage sites of tissue injury resulting from iron excess are: pancreas damage, cirrhosis, hepatic cancer, and gonad injury. Iron overload also presents with a tanning effect that result from its deposits in the skin. In addition, there is the emerging evidence that excess iron levels directly injure blood vessels. One likely mechanism for increased iron entry into the body is the type A personality type who is always in survival mode (the stress response). The stress response unleashes the primitive adaptation acute phase reactants. Ferritin is an acute phase reactant. Its increased presence provides a mechanism for why stressed owners (type A personality types) may accumulate more iron than is healthy.

Common nutrients and minerals need to be present in the diet for the availability of adequate molecular parts. The common hormone influences on what happens to these raw foods and minerals, once they are absorbed into the body, was discussed. The following discussion centers around how the different digestive juices dismantle the molecular building parts themselves. The body constantly needs reusable molecular components for regeneration and fuel. When raw food dismantles with precision there will be reusable components available for absorption.

Chapter 10
A Trip Down the Digestive Tube

With the completion of what is ideally contained in the diet, it is time for an imaginary trip down the digestive tube. Imagine shrinking down to the size of a single cholesterol molecule and hopping aboard an indigestible glass ship. The final destination will be the toilet bowel. The trip begins within a bite of food that contains proteins, carbohydrate, fats, vitamins, and minerals. A spinach and squash quiche would contain most of these.

The first thing most travelers notice would be the roughness of the ride beginning in the mouth. In the mouth: chewing up and down with the salivary juices that secrete into this one bite of food. The juices squirt out from various chambers of the digestive tube and contain precision food dismantling machines (enzymes). All of the digestive juices from each chamber prove very adept at disassembling the architectural framework of the different foods into their building block components.

The mouth is the first digestive chemical reaction chamber. In the mouth, the salivary juices contain enzymes. The salivary secreted digestive enzymes begin the dismantling of fat and carbohydrates contained in this bite of food.

The first juices secreted in the mouth contain machines (enzymes) capable of breaking carbohydrates down into the simple sugars. Additional enzyme machines dismantle fats into free fatty acids. The disassembly of carbohydrates occurs rapidly. The rate of disassembling fats, by these first machines, occurs rather slowly.

All human digestible carbohydrate breaks down into glucose, fructose, or galactose. Common examples of indigestible carbohydrate occurring in nature are wood and various plant fibers (lettuce). Simple sugars start to become available for absorption starting in the mouth.

In contrast, fat disassembly occurs as a much slower digestive break down process. One fat molecule is called a triglyceride molecule. One triglyceride molecule is made from three fatty acid molecules joined together by a single glycerol molecule. The act of digestion frees three fatty acids and one glycerol per triglyceride molecule. These dismantle in the process of digestion. Fat cannot absorb until it breaks down into these component parts.

Understanding What Fat is When it is Swallowed *(Optional Reading)*

The type of fat swallowed in the bite full of food is important. There are many different types of fatty acids found in nature. Different types of food break down into their own unique fatty acids. Before disassembling fat, the bite of foods fatty acid content has combined with glycerol. Each glycerol can connect to three fatty acids. The combination of three fatty acid types connects to one glycerol, which makes each type of fat (triglyceride). The combination of the different types of fatty acids makes up the fat content type in a particular food. Scientists call some fatty acids essential because the body cannot manufacture them. These essential fatty acids have powerful effects on the way blood vessels respond to injury, tissue inflammation, and the immune system response to various stimuli. An imbalance of essential fatty acids could tip the scales towards disease.

Fatty acids can be two carbons (acetate) to twenty-four carbons long. In this world, each carbon binds four times to the same or to different elements at a time. Each carbon atom is always bound to something four times. If two of these bonds bind twice to another carbon in the fatty acid chain, scientists call this an unsaturated fatty acid. If the double bond occurs only one time within a given fatty acid it is mono-unsaturated. If the double bond occurs more than one time in a fatty acid it is called polyunsaturated. The more unsaturated (the more times carbon binds twice to another carbon) in a fatty acid, two things occur.

First the fatty acid twists up in a bulky way making it more likely to be liquid at room temperature. Visualize this as the difference in gathering wood for a campfire when it is all bent and twisted versus the straight sticks. The twisted sticks are cumbersome and awkward. They don't stack well like straight sticks. Polyunsaturated fatty acids don't stack well. Because they do not stack well, they are liquid at room temperature.

Second, whenever a double bond occurs more than twice within a fatty acid, it confers a reactivity to rust promoters that may be in the blood stream. If the fatty acid is mono-saturated it reacts less readily to rust promoters than the poly-unsaturated varieties. The type of fatty acid content present, in the bite of food, determines stability or reactivity.

Returning to the Voyage Down the Digestive Tube

Back in the glass ship, note that up to 30% of the fat and a slight majority of carbohydrates in this bite of food digests by the saliva containing digestive machines (enzymes). Farther down the tube, different digestive machines release from the pancreatic secretions. These secretions aggressively continue to dismantle the remaining complex carbohydrates and fat molecules contained in this food.

Suddenly, a violent lurch forward occurs and the tiny glass ship moves into the esophagus, traveling at about four centimeters a second toward the stomach. Swallowing coordinates so that when the food and salivary juices reach the upper stomach valve it opens and permits smooth passage to the acid bath chamber.

Acid Bath Chamber

The stomach constitutes the second digestive chemical reaction chamber. In a healthy stomach miracles occur. Special cells deep within tunnel like pits open into the inside of the stomach. These special cells lining the numerous tunnels that exit on the stomach surface are called parietal cells. The parietal cells make and secrete stomach acid. A very good reason explains why these cells lay hidden beneath the inner surface of the stomach underneath the further protection of stomach mucus. The acid these cells secrete is so powerful that it would digest the acid producing cells themselves. Looking out of the glass ship in the stomach chamber, one would see that beneath the mucus layer there are numerous small pits that are poke-a-doted all along the middle area stomach lining. From these pits, acid juices would be flowing out from their source deep below.

The second type of stomach cell, chief cells, occurs deep down inside these tunnels, as well. These tunnels also open on the inner stomach surface. The chief cells specialize in the production of one model of protein disassembly machine. The originally secreted version of this enzyme machine for protein disassembly, pepsinogen, remain inactive until its 'wrapper' gets pulled off, whereupon scientists call it pepsin. The wrapper is composed of an amino acid chain that conceals the active part of this protein disassembly machine. To pull the wrapper off of pepsinogen requires that adequate acid be present or inactive factory product ends up lying around in the stomach doing nothing.

Adequate acid also needs to be present to provide working conditions that this model of protein dismantler requires. Further down the tube the enzymatic machines require the opposite working conditions. These enzyme machines need basic pH (alkaline) to activate. As soon as the food leaves the stomach there must be adequate alkaline juices flowing out of the pancreas or the next group of digestive machines will not activate. Different digestive chambers require different pH balance to activate the enzyme machines pertinent to that chamber. Some owners get into trouble because they have not been counseled on ensuring the proper pH balance for the chamber activities in question. In the stomach the pH needs to be sufficiently acid or the digestive enzyme machines will not activate.

Adequate stomach acid also protects owners from the passage of intact bacteria, viruses, fungi, and various digestible protein toxins. It protects because acid destroys these potential invaders. Some toxins do not adequately digest within the digestive tract. Inadequate digestion (disassembly of foreign toxins) can lead to food poisoning. Acid content creates an important first line defense against opportunistic pathogens. These pathogens wait for the chance to enter a body. Very few microorganisms can survive the acid bath in the stomach. This sterilizes the contents entering the small intestine under normal circumstances. If a microorganism makes it into the small intestine, it confronts many more noxious surprises. The stomach needs to make adequate acid.

Acid is needed for three things. First, acid destroys pathogens. Second, acid signals the stomach-esophagus valve to close tightly preventing heartburn. Third, acid needs to be present in sufficient amounts to stimulate the pancreas when the food exits the stomach. Only when the pancreas receives adequate acid stimulus will it vigorously release alkali and other powerful digestive machines that it manufactures.

Various informational substances (hormones) secrete into the blood stream and into the digestive tube at specific sites as food travels toward the rectum. The stomach chamber involves the first digestive chamber where hormones play a role in the strength of secretions. The message content of the hormone, gastrin, stimulates histamine. Histamine stimulates acid release. In a healthy stomach, the informational content released coordinates efficient dismantling and absorption of nutrients in a meal by the acid and enzyme secretion.

A cornerstone principle for healthy digestion concerns the ability to secrete adequate stomach acid. Paradoxically, many owners suffer from heartburn symptoms because they have a stomach acid deficiency. Additional owners suffer acid indigestion symptoms because they do not make enough protective mucus (see further discussion in this chapter). Until these situations are identified, which process (acid over production, deficient acid production or deficient mucus production or both) these owners's health will suffer.

Dr. James Privitera, author of *Silent Clots*, says it very well. "I don't know any body process that improves with age. Telling patients that as they age their stomach improves its acid output is inconsistent with this fact." Dr. Privitera realizes that acid deficiency or mucous deficiency often predates indigestion problems.

The minority of heartburn patients makes too much acid. In these cases, their stomach problems become amenable to the expensive acid suppressors available. Many patients are incorrectly diagnosed with acid over production when the real problem causing their heartburn results from acid under-production. Weak acid accessing the esophagus can still burn a hole and cause painful irritations. Acid does not belong in this anatomical area. Suppressing weak acid output damages other digestive processes. (*Wright and Gaby, 1998*)

The stimulus for the stomach-esophagus valve to shut tightly is the presence of adequate acid in the stomach contents. Without adequate acid to stimulate the tight closure of this valve, heartburn symptoms occur when a patient lays down with a full stomach. Physicians denote this condition as gastro-esophageal-reflux disease (GERD). The mechanism involving under production of acid, with consequent reflux backward through an incompetent stomach valve occurs in many heartburn sufferers. There are a few patients who have incompetent stomach valves for other reasons (hiatal hernia).

Another consequence of inadequate stomach acid secretion concerns the diminished pancreas stimulus to secrete its juices. The pancreas needs adequate acid to stimulate release of digestive and acid neutralizing juices. Without an adequate pancreatic stimulus, further protein and fat dismantling and absorption become compromised. This manifests clinically as patients who habitually avoid high protein meals because of the digestive difficulties that follow.

Acid deficient output diminishes digestion and absorption of critical minerals. Calcium, magnesium, iron, copper, zinc, and nickel all require acid in order for the operation of the 'one for one' acid exchange pump. The mineral exchange pump that the stomach is responsible for allows the absorption of the above needed minerals. These types of mineral absorption pumps need acid [H+] to exchange for each trace mineral absorbed.

Finally, adequate acid production provides powerful protection from the dirty outside world of microorganisms seeking access to the internal anatomy.

There are powerful alternatives available to the standard prescriptions used to suppress acid production in the treatment of heartburn. Healing involves soothing the inflamed tissue (esophagus, stomach, or duodenum). In the cases of stomach inflammation, there is likely a diminished mucus production. Mucous production depends on the presence of high quality hormonal fats (essential fatty acid derived) and adequate IGF-1 levels. The non-steroidal anti-inflammatory medication, like aspirin, poisons the hormonal fats. This causes a decrease in mucous production and an increase in stomach vulnerability towards irritation and erosion.

Owners with diminished IGF-1 (in the older literature called sulfation factor) lack the stimulus to form mucous. Stomach mucous is formed from GAG (glycosaminoglycans). Sulfation describes the critical step in GAG formation from the building blocks galactosamine and glucosamine where sulfate adds on. Adequate sulfate is required to keep mucous fully functional (slimy). It is the adequate interplay between IGF-1 and the hormonal fats (prostaglandins) that determine whether sufficient mucous production occurs. Rather than educate physicians and patients about these basic interrelationships they get groomed into thinking in the dumbed-down position. The dumbed down position occurs when basic scientific relationships are eviscerated from the medical textbooks or they are discussed in a fragmentary manner.

Certain plants possess the ability to calm and soothe irritated digestive tissues by creating mucous. They are not commonly acknowledged because they are affordable and effective. Licorice root proves very effective in soothing inflamed gastrointestinal tissues. Unprocessed licorice root can raise blood pressure by prolonging the influence of fluid retaining hormones made by the adrenals. This fraction of licorice root needs removal. This process is known as de-glycerrhization. Enzymatic Therapy Co. makes an excellent form of this root in a powdered form. If an owner takes this processed form when the next attack occurs, relief will be minutes away. The heartburn relief will occur without altering the gastrointestinal physiology. Licorice root powder creates thick mucous when swallowed with small amounts of water. This mucous is like the protective mucous made naturally by the stomach and protects against digestive acids. Unlike acid suppression pharmaceuticals, de-glycerrhized licorice root has no side effects and allows irritated tissues to heal.

After adequate healing of the inflamed tissues occurs, the next step can be undertaken. A holistically trained physician needs to supervise the next step because certain stomach conditions contraindicate this approach. The next step involves the supplementation of stomach acid with meals. The goal is to restore adequate acid to the digestive process in order to make nutrient digestion more effective. Deficient acid producers are often deficient in secretion of the first protein-dismantling machines called pepsin. It is beneficial to take a supplement that contains acid and pepsin. The product label will say pepsinogen denoting the inactive form. When combined with water and acid it will be activated to pepsin. With successful acid and pepsinogen supplementation, patients notice an increased tolerance for high protein meals and the ability to digest larger portions of steak, fish, and chicken, etc.

Typically, acid supplementation should be done during consumption of a protein meal. Adequate fluid should be swallowed with the hydrochloric acid pills, as well. The best dose per pill contains approximately 600mg of hydrochloric acid. The dose per mid-meal should be increased by one pill until a warm feeling is noticed. This sensation denotes that the correct dose of acid supplement has been exceeded by one pill. With the next meal decrease the dosage by one pill. If acid deficiency involves part of the digestive problem, supplementation will lead to feelings of increased well being following protein meals.

Alternatively, consuming alcohol with meals enhances the digestive process through the stimulation of acid output. Caffeine has been documented to stimulate acid secretion as well. Some of these fluids destroy the mucus lining that protects the stomach from digesting itself. Substances that disrupt the integrity of this mucous barrier are excess alcohol, vinegar, bile salts, and aspirin-like drugs (ibuprofen, aspirin, indomethacin, etc.).

The discussion of the hormones and nervous control of acid secretion can be complicated. Both of these acid producing mechanisms pathways converge on histamine, which increases when the vagus nerve activates and from the hormone, gastrin. Histamine directly causes the acid producing cells to release acid into the stomach. Blockers of histamine release (Tagamet, Zantac, and Pepsin) are commonly used to decrease acid release. The more powerful proton pump inhibitors (Prilosec and Prevacid) act by their ability to poison the acid producing cells in the stomach.

In summary, there are three main determinants of stomach function: the adequacy of acid production, the quality of the mucus layer which protects the stomach lining from digestion, and the quality of the enzyme machines which release when sufficient acid presents. Looking out from the glass ship into this stomach, all three of these processes occur in an orderly fashion.

The next stop in the glass ship occurs upon exiting the stomach through the pyloric valve. Immediately outside the stomach two drains that enter the small intestines are noticed. One drain comes from the gallbladder. The other drain comes from the pancreas. Food trickles out of the stomach valve. This causes each drain to gush with its own characteristic juice and mix with the partially digested food.

Pancreatic Juices and What They Need to Perform

The pancreas requires adequate stomach acid to stimulate it to release its stored juices. Pancreatic juices contain two basic components. The first component contains the acid neutralizing juice. It is needed because this chambers enzymatic machinery works only in an alkaline environment (the opposite of an acid environment). These digestive machines operate in the third digestive chamber, the small intestine. While the partially digested meal contents seep from the stomach, the pancreatic juices secrete. Pancreatic juices contain bicarbonate that reacts with the acid forming carbon dioxide gas and water. A healthy pancreas secretes more bicarbonate than there is acid present. The second component of pancreatic secretions, its digestive machines, requires an alkaline environment for proper function.

The second component of the pancreatic juice contains unique pancreatic enzymatic machines. Their design further dismantles fats, proteins, and, to a lesser degree, the few remaining complex carbohydrates. These juices comprise only part of the whole complement of digestive juices in the third chemical reaction chamber. The first chemical reaction chamber is in the mouth and the second in the stomach. Each of the chemical reaction chambers requires different work environments (acid or basic). The different work environments are necessary for the enzymatic machines secreted in that compartment to become fully functional.

Just like in the stomach, the enzymatic machines released within this digestive chamber, the duodenum, release with their packages wrapped around them. The enzyme machinery floats around idly and is useless until their wrapper is removed. The wrappers in this chamber are composed of amino acid chains that cover the active site of the protein disassembly machines. The inactivity precaution before secretion prevents the pancreas from digesting itself. The intestinal digestive chamber activation of the enzyme machinery depends on the cells lining the intestinal tube.

The cells that line the intestinal tube contain an unwrapping enzymatic machine. This particular machine activates only one of the pancreatic-produced enzymes. Scientists call this particular enzymatic machine trypsinogen while its still in the wrapper and trypsin when it unwraps. Enzyme machines produced in the intestinal lining cells unwrap trypsinogen to trypsin. The intestinal lining cells protect themselves from digestion by their own layer of mucus. Trypsin in turn unwraps all the other pancreatic digestive machines contained in 'wrappers'. Once trypsin has been freed, it begins to digest any protein that is not concealed behind the protective mucous barrier.

The Intestinal Chamber

The intestinal chamber is the third chemical reaction chamber. It receives chemical concoctions from the gallbladder and pancreas that dump into its proximal portion. The three different successive areas of this chamber reabsorb the majority of the secretions that occur higher up in the digestive tube. These chambers are the duodenum, the jejunum, and the ileum. The cells lining the intestinal tube look like shag carpet. One strand of carpet is composed of millions of intestinal lining cells called villus. Millions of strands make up the shag carpet and provide maximal absorptive contact with the digested food.

The intestinal lining is called the brush border where the digestive building blocks (amino acids, simple sugars, various fatty acids vitamins, and minerals) absorb into the body. Each type of molecular building component absorbs at specific sites along this digestive chamber. Like the stomach, the cells lining the small intestine tube protect themselves from digestion by a mucous layer. Each specific molecular part (amino acids, sugars, various fatty acids, minerals, and vitamins) has a specific transport method through this mucous layer. If a molecular part cannot successfully cross through this mucous barrier, it cannot assimilate.

The short life span of the digestive lining cells (the brush border) illustrates how body molecular parts are recycled and reused interchangeably. The recycling arises because these cell types continuously slough off into the digestive tube every two to five days. A snake sheds its skin in a similar fashion, but here it occurs from the inside of the tube. When these cells are sloughed off, they dismantle (are digested) into their molecular building parts. These mix with the food content contained in the digestive juices. These molecular parts then reabsorb by the digestive tract to be used somewhere else. Interchangeability of a molecular part ends only when it becomes damaged. Secretion of dead cells and the eventual re-absorption of the remaining useful molecular building blocks illustrate the interchangeability of the molecular parts that enters and leaves the tube. The interchangeability of molecular parts is an important concept.

Millions of recycled molecular parts move around the body. These same molecular parts build themselves into a structure that later dismantles into its molecular building parts only to again recycle into yet another molecular structure. This occurs constantly at different locations throughout the body. The end to the interchangeability of molecular body parts arises only when they become damaged. Replacement of the damaged molecular parts explains why a continuous supply of new quality molecular parts must become available through the diet and digestive assimilation.

In summary, the stomach allows seepage of its contents out of the pyloric valve and into the small intestine. Vigorous pancreatic contractions release sufficient acid neutralizing bicarbonate. Bicarbonate release occurs in excess of the acid present to produce the opposite (alkaline fluid) work environment. An alkaline pH optimizes the simultaneous release and activation of the unique pancreatic digestive machines. The pancreatic digestive machines are designed to work best in an alkaline environment. In the small intestine chamber the final dismantling process of fats, protein, carbohydrate and the genetic material building blocks occurs. The third digestive chamber also contains the sites where these molecular building blocks absorb into the body.

The Gallbladder Secretion

The gallbladder secretes juice out of the second hole in the proximal duodenum. Fat in the meal forces the gallbladder secretion to be a simultaneous secretion out of the second drain hole. Fats need additional molecular concoctions in the intestinal chemical reaction chamber in order to be dismantled. The gallbladder secretion solves the problem of fat floating on water. The meal content floats in copious digestive juices that are water based. They do not mix well with fat. The gallbladder secretes salts and acids that will break oil into tiny droplets. This process explains how the dairy industry makes the fat in commercially available milk stay dissolved in the milk. These substances are called emulsifying agents. The gallbladder secretes emulsifying agents made from acids of cholesterol and the break down of hemoglobin salts. In specific ratios, these substances raise the fat absorption from 50% without a gallbladder to 95% with a gallbladder.

The entire contents of the gallbladder secrete into the upper small intestine (duodenum) and are reabsorbed at the end of the small intestine (the ileum). After being reabsorbed, they quickly re-secrete into the gallbladder at such a rate that they recycle 6-8 times per day.

Nutritionally - The Owner is Only as Good as What He/She Absorbs

The healthy owner accomplishes the orderly uptake of different nutritional building blocks. Uptake becomes possible and occurs as individual parts become available to the brush border cells. Billions of brush border cells line the small intestine tube. In this digestive chamber, the quality of juices that come out of the pancreas and the gallbladder significantly effect how the body absorbs nutrients. This concerns the health of the brush border cells and their associated protective mucus.

The small intestine handles more than nine quarts of fluid per day. By the time meal remnants reach the large intestine (colon) only 1-2 liters of fluid remain. By the time they leave the colon in the bowel movement, less than one cup of the original nine quarts passes in the feces.

Digestion involves the secretion and re-absorption of large amounts of water. The content of the fluid changes considerably as the food moves down the digestive tube. These changes facilitate the dismantling tasks that need to occur. The fluid, salt, and even the enzymatic machinery parts largely reabsorb and recycle again and again.

The voyage down the small intestine in the glass ship has seen almost all the nutrients absorb that the meal contained. The meal remnants contain mostly indigestible fibers and friendly bacteria that comprise 50% of stool weight. Only small amounts of fat and protein still remain in the stool. The glass ship arrives at the passageway to the next digestive chamber, the cecum. The next digestive chamber is the large intestine (colon).

Health Cannot Occur Without a Happy Colon

The colon (large intestine) comprises the fourth chemical reaction chamber for the nutritional remnants still in the meal. In addition to finishing the digestive process, the colon is one of six organs that take out the trash (see *The Body Heals,* **Taking Out the Trash section**). Constipated people do not take out their colon trash very well. On the opposite extreme, are those owners who have chronic diarrhea. Chronic diarrhea means that the body discards important minerals, water, nutrients, and vitamins with the trash. On top of this backdrop of colon tasks concerns the unique requirements of the healthy colon being colonized by helpful bacteria.

The large intestine requires certain bacterial colonies to assist it in performing its many biological functions. If the right bacteria live in this chamber, good things happen. The right bacteria (mostly acidophilus and bifidus bacteria) re-acidify the meal remnants again. The meal remnants are all that is left of the meal that has finally made its way to the colon.

It is very important to overall body health that the colon contains ample friendly bacteria. These bacteria produce many B-vitamins and vitamin K. The colon-inhabiting bacteria feed the colon lining cells by changing indigestible fiber into carbohydrate. Second and more importantly, these bacteria change some of this carbohydrates into the short-chained fatty acids that are particularly nutritious for the needs of the colon lining cells.

If one of the colon functions isn't working correctly health consequences occur. It is worthwhile to consider some of these imbalances and the simple ways an owner can return to balance. The first imbalance concerns the colon's role as one of the organs responsible for preventing the fifth path to an old body, taking out the trash. It is important to identify the role of the colon in this path to longevity. The fifth path is about taking out the cellular trash that accumulates daily as the cells function doing their work.

When certain conditions exist, the colon becomes compromised in its ability to perform the necessary trash removal. The first compromise involves the wrong bacteria or amounts inhabiting the colon chamber. The second arises from trash removal problems that occur when the colon contents become fiber deficient. This prevents helpful bacteria from being able to create butyrate for the colon lining cells nutritional needs. Butyrate is a short chain fatty acid that colon cells prefer for energy. Third concerns the water retention role that fiber plays in keeping bowel movements soft. Fourth involves the chronic retention of feces or constipation that encourages the formation of toxins (putrefaction). Putrefaction results in the re-absorption of some of these poisons back into the body.

The last compromise concerns yeast organisms over growing in the colon chamber. Decreased acidity encourages yeast overgrowth. Sufficient colon acidity depends on adequate amounts of acid-producing bacteria. Whenever an owner takes antibiotics, the risk arises that these friendly bacteria will die. Friendly bacteria also die from chronic consumption of chlorinated water. When friendly bacteria die, the acidity of the colon decreases and this encourages yeast overgrowth. This allows the release of more toxins into the system. A careful, laboratory performed stool analysis can identify most of these common problems.

All of the trash removal problems are made worse when certain hormones become deficient in the colon. The primary hormones for colon health are thyroid, cortisol, IGF-1, and vitamin A. When any of these become deficient, colon health suffers.

Low thyroid function manifests as chronic constipation because thyroid message encourages energy production in the colon cells. Only through sufficient energy creation can the colon motility and activity be normal. Cortisol deficiency leads to colon inflammation and mucus abnormalities. Therefore, colitis is very responsive to cortisol medication. Often these patients have a thyroid problem as well.

Physician's Sidebar
Decreased IGF-1 levels lead to a decrease in the protective layer of glycosaminoglycans occurring throughout the body. The glycosaminoglycan layer protects the cells that line the body cavities (respiratory tract, gastrointestinal tract), blood vessels, the matrix that inflates and holds cells together within the body and the body surfaces ability to hold water and prevent shriveling (skin and organs). The IGF-1 level determines the adequacy of this protective layer in the colon. The consequence, within the colon, of the glycosaminoglycan layer becoming deficient concerns an increased propensity for toxic molecules to leak into the body. The increased toxic load further burdens the liver (see liver chapter).

Many physicians remain unaware of these important roles of IGF-1. Part of this results from the numerous different names given to IGF-1. In order to connect the overall effect of IGF-1 within the body a physician would need to know the other names for IGF-1 within the medical literature: nonsuppressible insulin like activity of the blood, sulfation factor, and somatomedin C. They would also need to know that IGF-1 levels depend on a healthy liver, adequate androgen (DHEA and testosterone) and sufficient growth hormone release. The sulfation factor nomenclature describes the fact that sufficient amounts of this hormone are needed to create the highly charged GAG's (glycosaminoglycans) that form in the above-described areas within the body. Sufficient amounts of GAG in the above areas allow sufficient barriers to protect body tissues. Deficient GAG in the colon sets up the conditions for leaky gut and colon inflammatory diseases. Deficient GAG in the blood vessels increases the overall injury rate to the underlying lipid bilayer. Deficient GAG within the cells and around the cells causes a dimunition of these electrically charged meshworks that hold water in the body. When water content decreases the shrinkage into old age begins.

Vitamin A makes cells grow up. Cancer is a problem of cells not growing up. Some cancer cells do not grow up solely because they are deficient in Vitamin A. Sufficient Vitamin A is necessary to instruct DNA programs. Vitamin A is particularly important in the same cell types for which the glycosaminoglycan layer forms a barrier. Signs of Vitamin A deficiency show up most obviously on the skin. Examples of this include roughened skin, pigmentation spots, and fine wrinkles. When these are present the same tendency exists in the colon. Colon cancer risk can be lowered by adequate Vitamin A intake. Vitamin A is found in high levels in carrots, squash and avocados.

The last chapter only skimmed the liver's role in whether one is fat or lean. Follow the fuel and one will understand many middle-aged diseases. Obesity, heart disease and diabetes, in their common forms begin in the liver. The liver determines what happens to fuel following a meal and between meals. The above-mentioned diseases begin because the liver misappropriates fuel.

Chapter 11

The Liver

Understanding the liver's role, in where fuel channels, facilitates a deeper appreciation for how obesity is only one symptom of many other diseases. These diseases often begin in middle age. They include heart disease, adult onset diabetes, high blood pressure and muscle wasting.

Healing from obesity involves helping one's liver to do the right thing. An overview of what other things the liver performs is also included in this chapter. After all obesity is only one symptom of the aging process. A sustained weight loss is not possible until healing occurs for the underlying cause. One of the big components of causality in the aging process concerns the liver misdirecting its available energy. This chapter explains how the liver can be encouraged to perform in the best interest of the body.

The liver serves multiple roles in the body. The tasks of the liver can be arranged into five main groups.

1. Disarms and removes toxins
2. Controls the availability of fuel in the blood stream

3. Facilitates the assimilation of fat into the body
4. Stores and releases minerals, vitamins and hormones
5. Manufactures and releases transport proteins into the blood serum.

Each of these five main liver tasks proves fundamental to health. The assessment for how the liver performs in these five main areas is often superficially addressed. Longevity requires that all five of these liver tasks perform at efficient rates. **In one way or another obesity has a component of causality in a sick liver.** Sick livers manifest from toxic injury (see subsection below), the wrong message content that instructs it to do inappropriate things with body fuel and from poor nutritional choices.

Disarm and Remove Toxins

The liver needs certain molecular parts to disarm many different types of toxins that are encountered when the owner creates or ingests them.

Toxins made by every day processes:

Natural body waste   Drugs
Ingested toxins   Cookware
Food additives   Food pollutants
Air pollutants   Water pollutants
Herbicides   Heavy metals
Toxins absorbed from an unhealthy colon

The liver has a choice on how it will disarm any toxin. There are five common mechanisms for the initial inactivation of toxins. In general, all five of these processes facilitate the next phase of liver detoxification. In the initial stage the liver machinery creates a molecular appendage, which it can use to attach the final removal compounds to. This allows the toxin to become water soluble or inactive. The final removal compounds cannot attach until one of five initial reactions occur (molecular appendage types).

The five initial liver deactivation methods are:
1. Oxidation

2. Hydrogen addition
3. Hydration (addition of water)
4. Cleavage through hydrolysis (removal of water)
5. Removal of chloride, fluoride, bromide or iodide

The trouble with these processes concerns the fact that they each tend to create reactive intermediates. The liver needs protection from these intermediates. The protective molecules needed by the liver are commonly called antioxidants or 'rust retardants'. The better the supply of these substances in one's liver, the more protection one has from liver injury. The greater the load of toxins, the more antioxidants needed. More antioxidants become necessary because they will be used up quickly.

The basic list of needed liver antioxidants is:

Vitamin A	Vitamin E
Vitamin C	Bioflavonoids (berries)
Selenium	Zinc
Coenzyme Q10	Pycnogenol (grape seeds)

Lipoic acid (real foods only)
Thiols that are found in garlic, onions, and cruciferous vegetables

These anti-oxidants are the basic protectors of the liver tissue. The liver cells need the protection of anti-oxidants because of the reactive intermediates created when the liver begins the first phase of deactivating and/or removing a body toxin. There are also certain vitamins and nutrients that are needed to power the molecular machinery that performs the task of disarmament.

The basic list of these vitamins is:

Vitamin B1	Vitamin B1
Vitamin B2	Lipoic acid
Co enzyme Q10	
Vitamin B3	Pantothenic acid
Folic acid	
Vitamin B6	Phospholipids like lecithin

The liver must begin the initial attachment of molecular appendages without releasing reactive molecules that can oxidize liver cells. The antioxidants prevent the oxidants from causing liver rust. These initial deactivation steps require specific vitamins to power the enzymatic machinery needed for these activities. Once the liver cell has created the various appendages on a toxin, it needs to proceed on to deactivation.

The basic choices for the final reaction in toxin removal are the addition of:

- Sulfate
- Glucuronic acid
- Glutathione or n-acetyl cysteine
- Acetate
- Methyl
- Certain amino acids: glycine, taurine, glutamine, ornithine, and arginine

Depending on the final solubility characteristics, the deactivated toxin will either be excreted in the bile or blood stream. When excreted in the bile it will be removed in the feces. However, poor bacteria content in the colon can prevent this and the toxin can be absorbed back into the body. Some toxins reabsorb because the wrong bacteria rip apart the deactivation appendage that the liver attached.

When a water-soluble toxin excretes into the blood stream, it heads for the kidneys that can remove large amounts of toxins. Many toxins and hormone excesses that the kidney removes must first be made water-soluble by the liver. This includes ammonia, steroids, small chains of amino acids, and heavy metal complexes.

The liver exemplifies one tissue that possesses remarkable capabilities for regenerating new cells until its underlying architecture disrupts. Scientists call disruption in the architecture state of the liver, cirrhosis. Healing compromises because cell regeneration disorganizes. At the level of cirrhosis the energy template of where cells belong disrupts. When the energy template disrupts a progressive cellular disorganization ensues.

The liver is responsible for the up take of ammonia. Ammonia production generates from amino acid breakdown when protein converts to sugar (gluconeogenesis). The ammonia formed in the breakdown process converts to urea in the liver and exits in the urine. The kidney has the ability to excrete a limited amount of ammonia. However, it is the liver that neutralizes the majority of this toxin. Once the liver neutralizes a toxin it uses the bile or the kidneys to remove them (many environmental toxins and prescription drugs). The liver requires numerous nutritional factors to effectively remove toxins and this ability greatly diminishes without sufficient nutrition.

A rich system of blood vessels (liver sinusoids) exists in the liver. These arrange to allow immune system scavenger cells room to grab unwanted material out of the blood stream. They are fed directly from the portal vein that is the drain for all the blood in the intestines and colon. These concepts are discussed further in *Clinical Nutrition: A Functional Approach* by Jeffery Bland, PhD.

The Liver Determines Fuel Availability

Inappropriate fuel types in the blood stream cause diseases like diabetes, heart disease, strokes, and peripheral vascular disease. These diseases can have their origins in the liver. When the liver treats fuel types inappropriately, the blood vessels begin to break down. There are two common examples of inappropriate fuel types released by the liver with consequent blood vessel injury. First concerns the high blood sugar of diabetes. The second mechanism involves the high rate of liver synthesis and release into the blood stream of LDL cholesterol found in heart disease and one type of diabetes. In both cases excesses of these fuels injure the blood vessels inside walls. In both cases the liver creates and releases these fuels inappropriately and this causes the excesses found in these diseases. The cause of inappropriate release of fuel excesses results from the liver receiving the wrong hormones messages that direct it incorrectly.

Healing involves attention to lifestyles, adequate nutrient intake, and hormone balance at the level of the liver. Instead, owners are often told the hopeless mantra about the cruel hand that genetics has dealt them. The typical approach sells drugs and procedures. Healing involves improving the types

and amounts of hormones that instruct the liver on whether to store or release fuel.

Like other organs, the liver dutifully follows the message content it receives. Healing involves an assessment of the proportions of the hormones that deliver message content at the level of the liver cell. Only when the proper amounts of hormones instruct the liver will these diseases begin to heal.

The overall hormone message delivered to the liver determines whether the liver takes fuel out or puts fuel into the blood stream. Five main hormones determine how the liver directs fuel - insulin, glucagon, growth hormone, cortisol, and epinephrine. The message content arriving at the liver determines whether the liver will manufacture fuel in storage forms (glycogen and LDL cholesterol) or release them as readily combustible types for use in the cellular power plants (mitochondria). If the liver is in storage mode, it does this by removing fuel from the blood stream. The liver synthesizes LDL cholesterol particles with storage in mind. Even though the liver eventually releases them, their design is such that they head for the storage destinations: the macrophages that line the arteries and the fat cells.

The opposite situation occurs when the liver releases sugar and fatty acids into the blood stream. The sugar released readily accesses the cells when adequate IGF-1 is present in the blood stream. Once inside the cells this fuel either combusts in its power plants or serves for structural components in the cell. Four out of five of the above hormones encourage this type of fuel release.

It helps to arrange the hormones interacting with the liver in the antique weight scale analogy. This arrangement reveals that only insulin falls on the side of the scale tipping it in the direction of storing fuels (glycogen and LDL cholesterol). The other four hormones (at the level of the liver) counter storage and encourage the power plant accessible types of fuel release by the liver into the blood stream.

The insulin predominant hormone situation allows a tremendous increase in fat and cholesterol because insulin behaves like the liver's fuel tank nozzle connection. It preferentially desires to fill the liver cell fuel tanks because of its initial site of release, the pancreas, dumps it straight into the liver via the protal vein. This anatomical fact means that whatever the insulin secretion rate the liver will always receive the highest amount of insulin message content. When the liver cell fuel tanks fill up with carbohydrate (about 400 grams when adequate potassium remains available), insulin instructs the liver to make the extra sugar into fat and cholesterol. This explains why one needs insulin for fat cell growth and in the storage of muscle and liver glycogen.

Counter regulatory hormones can overpower insulin's ability to direct carbohydrate and fat storage in the liver (each to varying degrees). Counter regulatory hormones direct the liver to dump stored fuel into the blood stream for usage in cell power plants.

When between meals, mentally or physically stressed the counter regulatory hormones levels become elevated. The fat stores are needed during intense and prolonged exercise. Exercising muscles and heart prefer fat as their fuel source. Carbohydrate storage is limited to about 1500 calories in the liver and about 500 calories in the muscles. When the glycogen stores are burned, as in the case of prolonged exercise, the body needs other fuel sources for energy. A 150 pound athlete with 15% body fat has access to 22.5 pounds of fat times 3500 calories per pound of fat (about 80,000 calories). The proper counter regulatory hormones allow access to the tremendous fat fuel storehouses.

Nerves and red blood cells exclusively need sugar for all their energy needs. This fact explains the desirability of adequate counter hormones that direct the additional release of fat. Processes that conserve glycogen for the nerve and red blood cells enhance endurance because nerve and red blood cell's optimal function are preserved when their sugar supply remains available. This happens when the muscles and heart have access to more fat for their energy needs. Conversely, exercising owners that insist on promoting high insulin states compromise their red blood cells and nerves due to sugar stores being consumed more quickly. When insulin excess occurs in the liver, it competes with the ability of the liver to dump fat into the blood stream and create more sugar from amino acids. The sugar stores dry up with the same level of activity because less fat is available. Exhausted sugar stores in the exercising athlete are commonly known as "the wall." The wall will arrive more quickly when insulin levels are high.

Longevity and exercise performance critically relate to the balance of hormones that instruct the liver. When these hormones improve not only exercise ability but also heart disease, blood vessel disease and diabetes risk and complications diminish as well.

Insulin Causes Fuel Storage as Fat or Glycogen

The counter hormones, glucagon, growth hormone, cortisol, and epinephrine all cause the liver to release and synthesize readily accessible fuels into the blood stream.

Initially, insulin will only direct the liver to remove sugar from the blood stream for storage as glycogen (if sufficient potassium remains available). It can store about four hundred grams of sugar in this manner. Sedentary owners have livers that are almost always full of stored sugar. In this case, insulin directs the liver to make the excess sugar into fat and cholesterol that are packaged for the fat depots. These depots are in the fat cells and in the macrophages that line the arteries. At the level of the liver, insulin is the only hormone that promotes storage of fuel. The other four hormones oppose the message content of insulin at the level of the liver.

Glucagon counters insulin by four ways in the liver. First it stimulates the liver to change available amino acids into sugar (gluconeogenesis). Second, it stimulates the release of stored sugar (glycogen) from the liver into the blood stream. Third, it stimulates the release of stored liver fat into the blood stream. Glucagon's message content makes body fuel available. Fourth, the glucagon message content decreases cholesterol and fat synthesis from carbohydrate. When cholesterol and fat are manufactured at slower rates, LDL cholesterol levels in the blood stream will go down.

Growth hormone has some of the same effect on the liver as glucagon does. **There are two important exceptions**. It inhibits the conversion of amino acids into fuel. This is known as a protein sparing effect. When growth hormone levels are high, it counters the ability of glucagon to convert protein into sugar. The advantage of having high growth hormone levels relative to the other counter regulatory hormones is that it conserves protein. This fact is important for athletes because muscles are made from protein. In contrast, all other counter regulatory hormones to insulin are catabolic toward protein stores.

The second different effect of growth hormone's message content regards its ability to direct the liver to release insulin-like growth factor (IGF-1). When this hormone is outside the liver, it acts like insulin by facilitating the muscle cells to take up fuel (sugar). Remember that in healthy owners insulin-like growth factor (IGF-1) is found at levels greater than 100 times that of insulin in the blood stream. This liver secreted hormone reduces the need for insulin and does not stimulate the liver to make fat and cholesterol. A longevity advantage occurs when IGF-1 levels are high. High IGF-1 levels depend on three things: an adequate release of growth hormone, a healthy liver capable of high rates of IGF-1 manufacture, and adequate DHEA levels to stimulate the liver cell DNA programs to direct the manufacture of IGF-1.

Growth hormone's effects on the body need to be understood in a tandem-like fashion. Once the tandem of growth hormone release followed by IGF-1 release is recognized one can avoid the confusion discussed in the scientific literature. Medical physiology textbooks describe growth hormone as a diabetogenic hormone. This is a half-truth except when the liver is diseased or growth hormone production becomes abnormally high. Abnormal increases in growth hormone levels can occur with the disease acromegaly and with high dose growth hormone replacement therapy. However, healthy livers with adequate androgen message content promptly release IGF-1, sugar and fat when growth hormone levels rise within normal parameters. The

IGF-1 released lowers the amount of insulin required to bring fuel into the cells. Cortisol is the next counter hormone to insulin's message content. Cortisol powerfully counters insulin in its ability to remove sugar molecules from the blood stream. Cortisol directs the liver to release stored liver sugar and fat into the blood stream. It also instructs the fat cells to release fatty acid fuels into the blood stream. In addition, it acts on the protein stores of the body and causes them to release amino acids into the blood stream (catabolism). The liver then sucks up the released amino acids for processing into sugar. The process of gluconeogenesis denotes the conversion of liver-sequestered amino acids into sugar.

Physician's Sidebar

Normally, growth hormone will counter a diabetes tendency by its message that directs the release of IGF-1. IGF-1 behaves like insulin in the circulation. This happens because of the tandem effects of growth hormone release followed by the liver simultaneously releasing IGF-1, sugar and fat. Medical physiology textbooks focus on the fact that growth hormone initially causes the release of sugar into the blood stream. However, the tandem of IGF-1 release that normally follows causes the peripheral uptake of sugar out of the blood stream (i.e. outside the liver). It is the tandem of growth hormone release followed by IGF-1 release that prevents the overall rise in blood sugar. The insulin-like effect of IGF-1 explains why growth hormone, in normal amounts, lowers insulin requirements in patients who have a normal liver and DHEA and/or testosterone level. It also explains the observation why normal DHEA levels lower insulin requirements. Insulin requirements lower because normal DHEA levels instruct the liver DNA programs to increase the amount of IGF-1 manufactured. IGF-1 releases when growth hormone levels elevate. The IGF-1 released, acts like insulin in the cells outside of the liver (the periphery). In the periphery, IGF-1 directs many cells to take sugar out of the blood stream. When adequate insulin-like growth factor secretes,

223

the insulin needed to normalize the blood sugar decreases. Some physicians feel that insulin, outside of the liver in the healthful state, is unnecessary beyond very low levels.

IGF-1 has the additional benefit of facilitating the uptake of cell nutrition and minerals. IGF-1 occurs at 100 times greater the amount of insulin in the circulation when an owner is healthy. This makes sense since there is about 100 times more insulin like activity needed to direct the proper nourishment of the cells outside the liver and fat. The highest pure insulin type receptor concentration occurs on the liver and fat cells at about 200,000 receptors per cell. Whatever the insulin release rate because of the anatomy of the pancreas secreting into the portal vein straight into the liver, the liver always receives the highest message content of insulin.

High levels of IGF-1 diminish the need for insulin in the cells outside the liver and fat. Conversely, low levels of IGF-1 require an increased insulin secretion because a sugar load will require more insulin to return the blood sugar to normal following a carbohydrate meal. However, healthy owners have more than 100 times IGF-1 in their blood stream as compared to insulin. Unhealthy owners have liver strain because more insulin forces the liver into sucking excessive sugar out of the blood stream. The extra insulin also increases the rate of LDL cholesterol synthesis. The imbalance between insulin levels and IGF-1 levels creates a situation that favors increased body fat formation

Extremely high cortisol release tends to deplete the body protein content. The metabolically active component of the body consists of protein. Diminished protein content logically extends to the need for fewer calories per day. It also causes smaller muscles and organ size because this is where the protein comes from. High cortisol release occurs with stress and when growth hormone levels have fallen. Remember, only growth hormone can protect body protein content during times of stress (fasting, exercising, between meals, worry, etc.)

Physician's Sidebar

Cortisol instructs the liver cell DNA. The liver cell DNA, when instructed by cortisol, leads to the formation of certain cell receptors. The cell receptors are needed to recognize hormones that are not as powerful as level one-type hormones. Only level one-type hormones, like cortisol, can directly instruct the DNA programs. The cortisol directed receptors are necessary to recognize adrenalin (epinephrine) and glucagon. These are level three and two hormones respectively (see the hierarchy of hormones, appendix A). The liver will be unable to recognize the message content of epinephrine and glucagon without sufficient cortisol. Cortisol deficiency sets up a situation of the 'lesser hormones' not being able to deliver their message content (elevate the blood sugar in this case). The only other counter regulatory hormone to insulin, growth hormone, will cause the release of IGF that exacerbates low blood sugars when the liver fails to release enough sugar between meals. Here lies a little recognized mechanism for how low blood sugar occurs between meals.

Low blood sugars (hypoglycemia) are the hallmark of diminished adrenal reserve (see *The Body Heals*, Adrenal chapter). The adrenal gland manufactures all body cortisol. Deficient cortisol in the setting of high insulin will lead to diminished blood sugar levels. Unfortunately, many of these hypoglycemic prone owners are prescribed frequent feedings. This approach leads to weight gain and a compounding of problems with the passage of time.

However, a **24-hour urine test** will document diminished adrenal cortisol production. Occasionally, the additional test of Cortrosyn challenge (beyond the level of this discussion) of the adrenal reserve function will be needed to elucidate borderline cortisol reserve. There are several considerations for healing from diminished adrenal function. First, dietary changes for correcting carbohydrate and mineral imbalances are a strong start. Second, the results of the 24-hour urine test document the scope of the steroid imbalance that includes cortisol deficiency. In some cases replacement with real steroid hormones becomes necessary. In others, diet and life style changes suffice. Customizing replacements to actual needs prevents side effects from cortisol replacement. In addition, changing the carbohydrate and mineral content of the diet prevents side effects. The replacement program often includes other adrenal steroids as indicated by the 24-hour urine test. When these factors are considered together, side effects are reduced or eliminated.

Epinephrine is one other counter regulatory hormone that opposes insulin. Epinephrine acts like cortisol in its ability to free fatty acids from the fat stores. However, its release powerfully inhibits insulin secretion. Ephedra acts in a similar manner to epinephrine. This fact explains its weight loss effect. Blood pressure elevations from either one are unusual. In contrast, norepinephrine powerfully raises blood pressure. The more effective a natural remedy the more it gets bashed in the media by fear tactics. Aspirin like medications kill more people in one week than have probably ever been harmed by ephedra. Yet, aspirin can be bought over the counter and ephedra is on its way to being outlawed.

The other counter regulatory hormone, growth hormone acts mostly in the liver in its opposition to insulin. However, in one important way both insulin and growth hormone share a common message theme. These two hormones both have in common the conservation of body protein. Insulin conserves body protein following meals while growth hormone conserves body protein between meals (fasting and exercising).

The Liver Controls the Body Fat Content

Diabetes and heart disease are often diseases that are secondary to the liver receiving poor informational direction. 90 percent of all diabetics have an excess of total body insulin. Most heart disease owners also have an excess message content from insulin (increased insulin release rates). When insulin is the cause, these diseases, begin with poor message content at the level of the liver. The poor message content arises from excess insulin. The liver cell dutifully follows the message content it receives. Insulin in the liver opposes growth hormone, epinephrine, cortisol, and glucagon message content. The balance of message content between the hormone insulin and the other four counter hormones comprises a major determinant in the risk of an owner for developing these two diseases.

The complete cholesterol profile that is typically performed on an annual basis provides clues as to the summation of message content in one's liver. Some owners may have a normal blood cholesterol profile, but an obese body. **Only 50 percent of heart disease victims have abnormal cholesterol. The other fifty percent have normal cholesterol.** The fact that most heart disease owners are obese points to the fact that they have too much insulin because it is always a pre-requisite for body fat (see Syndrome X). With more insulin there is more message content to create and maintain fat and cholesterol. High insulin leads to stuffed macrophages that line the inside of arteries with LDL cholesterol. Fat laden macrophages eventually plug an artery even if the blood cholesterol is normal.

Whenever the lab results are sub-optimal or the owner is obese, several things need to be considered and investigated. With this and some life style changes, the excess insulin caused diseases are often effectively curtailed. Altering the message content that the liver receives proves critical. When these primary factors are ignored symptom control medicine becomes all that is possible.

Healing from heart disease or diabetes involves alleviating one or more of these message abnormalities at the level of the liver:

1. Unfavorable insulin and glucagon ratio of message content in the liver
2. Potassium deficiency leads to insulin resistance connection
3. Growth hormone connection
    a. Excess
    b. Deficiency
4. Cortisol excess
    a. Normal pancreas
    b. Wounded pancreas
5. Thyroid status

6. Growth hormone when the liver is unhealthy
7. Excessive proinsulin levels
8. Excessive estrogen levels
9. Low androgen levels

A key difference occurs between the disease processes of diabetes and heart disease. The difference concerns the elevated blood sugar within the diabetic. However, these two diseases share a common pathology. 90 percent of diabetics have elevated insulin secretion rates and the majority of atherosclerotic heart disease involves elevated insulin secretion rates, as well. In most adult onset diabetics, their body fat has finally elevated to the point (genetically determined) where their pancreas can no longer handle the increased insulin need and the blood sugar rises.

Five Basic Reasons for Elevated Insulin Need

Insulin need rises when body mass goes up, IGF-1 levels fall, total body potassium decreases, chronic stress predominates and/or carbohydrate consumption increases. When a pancreas cannot meet the insulin demanded by the excessive presence of any one or more of the above conditions, the blood sugar will rise. By definition, when blood sugar rises beyond 140, diabetes is diagnosed. However, the same process of elevated insulin plugs up the blood vessels of the typical heart diseased owner. The pure heart diseased owner, who is not diabetic, can still make enough insulin to keep his blood sugar normal. **Here lies the common fundamental common link between adult onset diabetics and most atherosclerotic heart disease victims: elevated insulin secretion rates.**

Many heart disease patients start out with normal blood sugars and only after many years their pancreas becomes exhausted. This commonality between these two diseases explains why mainstream medicine is finally acknowledging that at the time of diagnosis of adult onset diabetes, heart disease is presumed to already exist. **Ninety percent of diabetes and the majority of heart diseased owners have elevated blood insulin secretion rates. The high insulin levels that occur in both diseases lead to increased blood fat abnormalities.**

However, in the case of the adult onset diabetic, the amount of insulin needed exceeds the ability of their pancreas to excrete enough insulin. Adult onset diabetes results, when any one or more of the above five processes causes the pancreas to exceed its ability to excrete enough insulin. In all five situations, higher insulin output is required (insulin resistance). High insulin levels direct the liver to make excess LDL cholesterol. The excess LDL cholesterol released into the blood stream causes the vessel walls to collect LDL cholesterol in the macrophages. The elevated blood sugar of diabetics is an additional mechanism that injures the blood vessel.

Mainstream medicine tends to focus on normalizing the blood sugar with the price paid in the form of health consequences from the increasing insulin message content. Conventional physicians are taught to peripherally address the increased insulin caused, cholesterol and fat making side effects, with cholesterol lowering drugs. They are also taught to prescribe blood pressure medication for the high blood pressure caused by excess insulin or mineral imbalance (see *The Body Heals,* The Blood Pressure chapter). Healing requires that insulin levels come down rather than pummel the liver with poison.

That insulin feeds fat formation in ones arteries is particularly disturbing when one realizes how mainstream medicine treats many adult onset diabetics. They sidestep the central determinant, insulin excess. The majority of diabetics are treated by methodologies that raise insulin levels even further. This may normalize blood sugars, but it will only make the fat accumulating in the arteries worse. Clinically the higher insulin levels show up as a worsening cholesterol profile. The cholesterol profile worsens with these treatment strategies because the higher the insulin the more the liver receives message content to make cholesterol and fat from the sugar taken out of the blood stream.

Mainstream physicians are taught to remedy the worsening cholesterol profiles with the common class of liver poison, the statin drugs. Statin drugs inhibit the cholesterol making enzyme, HMG CoA reductase. Biochemistry textbooks discuss the fact that if the insulin message content decreases, this enzyme turns down its activity level naturally. In addition, the enzyme HMG Co A reductase turns down its activity further when glucagon levels within the liver elevate (see below). Rather than educate physicians about this buried and ignored fact the symptom control approaches provide the standard of care. These common symptom control approaches are always accompanied by side effects and have nothing to do with how one heals.

Physician's Sidebar

Another common symptom control approach to treating adult onset diabetes involves the metaformin like drugs given to increase insulin sensitivity. Their most common side effect, metabolic acidosis, provides a clue for how the increase in insulin sensitivity comes about.

Red blood cells and exercising muscles provide reliable sources for the constant creation of lactic acid. Normally, the liver removes this to keep the blood pH balanced. Anything that poisons the liver will decrease the liver's ability to perform this important task. In addition, a decreased conversion of lactic acid to pyruvate, a sugar intermediary, will lower the gluconeogenesis rate as well. A lowered gluconeogenesis rate will decrease the rate at which the liver can dump sugar into the blood stream and lower the amount of insulin needed. This is probably not the primary effect for how these drugs lower insulin requirements.

Remember, that the more IGF-1 available the less insulin required. Also remember that the bulk of IGF-1 binds to a carrier protein called insulin like growth factor binding protein (IGFBP). Associated with this complex of IGF-1 and its binding protein is the acid labile subunit (ALS). A rise in acid content within the blood stream will allow ALS to undergo a conformational change, which frees up and allows IGF-1 to help insulin with fuel delivery (an increase in insulin sensitivity).

An additional insight occurs that regards the natural process of increases in insulin sensitivity when the blood turns acidic. It has long been recognized that anaerobic exercise (weight lifting) raises growth hormone levels many more times than does aerobic exercise. Here again it is the rise in lactic acid levels that seem to increase growth hormone secretion. Drugs like metaformin tend to raise lactic acid levels, as they pummel the liver, and this fact recreates the rise in lactic acid seen with anaerobic exercise. As growth hormone levels rise the secretion rate of IGF-1 rises as well. In addition, the IGF-1 within the blood stream becomes more helpful with sugar removal because the ALS fragment has undergone a conformational change. Rather than the mainstream textbooks and eduction system helping physicians to understand the important roles that growth hormone and IGF-1 have in insulin sensitivity, they are given piecemeal information. Piecemeal-educated physicians enable drug sales to increase but the overall healing rate decreases.

Passive patients require these symptom control approaches but what about those owners who desire healing? The first step in healing an insulin-excess disease is to understand which type an owner may have. Contrary to the mainstream medicine approach to adult onset diabetes numerous additional causes for this disease process exist. Before healing can occur the afflicted needs to know which process causes his/her disease.

All of these nine factors affect the message content at the level of the liver by their unique effect on the amount of insulin message content present. **Remember that obesity has a powerful association with both heart disease and diabetes. The nine groupings below help explain why. When one understands why they have a disease, they can begin to heal. Healing is facilitated when one's physician inquires into which type of liver message abnormality promotes a given owners disease process.**

Insulin and Glucagon Message Content in the Liver

It is more accurate to say the activity of the cholesterol-manufacturing enzyme in the liver, HMG Co A reductase, is determined by the ratio, in amount, between glucagon and insulin. A high protein/good fat and low carbohydrate diet will tend to increase glucagon levels relative to insulin. Similarly, stress management and regular exercise will tend to improve this ratio (see exercise and muscles). Owners, who eat steak and eggs while simultaneously decreasing carbohydrates, will tend to see a drop in their cholesterol (see digestion of fat). There are other factors that influence blood cholesterol level, but this ratio is all-important in most owners. Without attention appropriately placed on an improved insulin and glucagon ratio healing a blood cholesterol problem becomes difficult.

Diminished Potassium Leads to an Increased Insulin Need

It was explained earlier that a fall in total body potassium leads to insulin resistance (more insulin is needed for the same sugar load). The increase in insulin means that more fat and cholesterol-making message will then be present in the liver. Diminished body potassium has detrimental effects to the insulin and glucagon ratio. This mechanism of altered blood fat creation from the liver is rarely recognized in America today. This basic defect arises because owners who subsist on a processed food diet stuff their cells chronically with the wrong mineral ratios. The body was designed to intake at least three times as much potassium as sodium. Processed foods contain a reversal of this natural mineral requirement (see mineral table). Around middle age the kidney begins to falter in its ability to conserve potassium and excrete excess sodium.

The potassium deficiency of middle age means less of this mineral becomes available to bring the blood sugar down following a carbohydrate meal. A single potassium is necessary to bring one sugar molecule inside a liver cell for sugar storage purposes. This is also true for all cells except in the brain and red blood cells. Inadequate potassium causes the pancreas to secrete extra insulin. Eventually it senses the delay in the blood sugar fall after eating. This will be met with resistance because the basic defect is not enough potassium to uptake the sugar into the liver cells. It is slightly more complicated than this because of the IGF-1 that is released between meals. Potassium deficiency impedes the storage of sugar as glycogen because a fixed amount of potassium per gram of glycogen in the liver needs to be present. The liver manufacture of fat and cholesterol does not depend on potassium levels.

The extra insulin that occurs in these situations will activate the fat making machinery in the liver preferentially. This happens whenever the liver receives excess insulin message content. In contrast, the healthy body has ample potassium to facilitate the rapid liver uptake of sugar out of the blood stream following a meal and to make glycogen. It also has sufficient IGF-1 to facilitate sufficient organ and muscle uptake of sugar that lessens the liver proportion of sugar removed out of the blood stream. Between meals or with exercise the liver stored potassium and IGF secrete together. With adequate potassium, IGF-1 can promote the uptake of nutrients. Potassium and other nutrients are needed for cell energy and rejuvenation (i.e. strong, large muscles). With insufficient potassium, the blood sugar normalizes more by the liver than by the peripheral cells (hungry cell syndrome).

The dependency of insulin on adequate body potassium exposes a defect in the high protein diet proponents. It clarifies how the fruit and vegetable proponents have a piece of the puzzle that pertains to diets that work. When both of these pieces correctly re-unite, a superior diet and cholesterol profile becomes possible.

The high protein diet premise improves when adding in the optimal ratio of potassium to sodium mineral content. This drastically lowers the insulin message content needed by the body. A lower message content of insulin will improve the ratio between it and glucagon. This will allow less cholesterol and fat synthesis in the liver. Diminished cholesterol and fat synthesis leads to improved blood cholesterol. When less insulin becomes necessary the pancreas strain diminishes. In the case of diabetes, the need for extra insulin becomes less likely.

Growth Hormone Connection - What Message the Liver Receives and its Effects on Diabetes and Heart Disease

Growth Hormone Excess

Excessive growth hormone tends to promote diabetes when its level exceeds the ability of the liver to secrete the consequent insulin-like growth factor [IGF-1]. Within the liver it promotes the release of sugar and fat into the blood stream. The second effect concerns the simultaneous release of insulin-like growth factor (IGF-1). IGF-1 acts like insulin out in the periphery, muscle, and organ cells. IGF-1 helps insulin take sugar out of the blood stream. The diseased liver cannot release sufficient IGF-1. This defect occurs at varying degrees of severity. Elevated GH with a diminished IGF-1 release will cause the blood sugar to rise. When the blood sugar rises the pancreas attempts to normalize it by secreting insulin. When blood sugar rises beyond the pancreases' ability to manufacture the insulin necessary to counter act the excess growth hormone, diabetes results. Growth hormone excess can occur either in acromegaly or with high dose growth hormone replacement protocols. It is the increased insulin, in the above scenario that leads to altered blood cholesterol profiles. When this occurs heart disease and eventually diabetes become more likely.

Within this subgroup of owners that may get diabetes are those with a wounded pancreas. Their High IGF-1 levels are their only protection from full on diabetes. Sometime in adulthood as the IGF-1 levels fall these owners develop diabetes. They can usually be recognized clinically as those atypical looking diabetics that have large muscles and diminished body fat for their age. This is a rare type of physique for any diabetic and should alert the clinician to look into the fasting insulin, C-peptide and IGF-1 level. Also serum iron studies identify those owners whose pancreas was damaged by its accumulation. Again these rare types of diabetic patients have as their fundamental defect a wounded pancreas and as their salvation the particularly strong ability to make IGF-1.

This type of diabetic patient benefits from insulin treatment because insulin deficiency, and not IGF-1 deficiency drives their disease process. In contrast, the typical emaciated insulin dependent diabetic usually suffers from IGF-1 deficiency that eventually exhausts his/her pancreas (beta cell burnout) and diabetes develops. Measures that raise IGF-1 early in the disease process could save some of these owner's pancreases from total burn out. Physicians would first have to know to check a fasting IGF-1 level. Paradoxically, this is not usually part of the mainstream education content.

Growth Hormone Deficiency

For different reasons, problems with blood sugar control occur in this situation. Blood sugar will tend to rise because less liver direction occurs to release insulin like growth factor (IGF-1). IGF-1 levels occur at levels greater than 100 times those of insulin in the blood stream of a healthy owner. IGF-1 helps insulin with taking nutrition out of the blood stream and into the cells. With less IGF-1, more insulin must be secreted for the same sugar load. The increased demand for insulin can exhaust the pancreas of some owners and diabetes develops. Sedentary lifestyles promote decline in growth hormone secretion rates. Conversely, owners that exercise, have higher growth hormone levels. That means less insulin will be required because of the increased liver stimulation to secrete IGF-1. Less insulin creates less message content in the liver to make fat and cholesterol. Athletic training will decrease cholesterol levels for these reasons.

Clinically, growth hormone deficient owners express as one type of syndrome X patient. Their fundamental health defect results from a fall off in their IGF-1 levels. Healthy people have over 100 times the IGF-1 compared to insulin within their blood streams. The liver is large and the pancreas is small. The smaller pancreas, when forced to increase its insulin production rate many fold eventually exhausts itself. A small fall in IGF-1 levels requires a marked increase in insulin production to offset the deficiency in the total amount of fuel nozzle hormones for the body cells outside the liver and fat. IGF-1 levels fall from three basic mechanisms: liver injury, decreased growth hormone secretion rates and diminished DHEA and testosterone levels, which direct the liver DNA to make IGF-1. Growth hormone tells the liver to secrete IGF-1.

Excess Cortisol Levels - The Message the Liver Receives Depends on the Health of the Pancreas

Excess cortisol creates obesity and/or borderline diabetes. In either case the blood vessels sustain damage. The phenotype distinguishes which process occurs. The 24-hour urine test confirms the physical exam and blood findings. With increased cortisol, owners with a normal pancreas, tend to become obese. With the passage of time, they will tend to appear cushingoid (moon face, buffalo hump, stria on the abdomen, panus formation on the abdomen, muscle loss). If the cortisol level elevates high enough, they can reach the upper limit of their pancreas for insulin production. Diabetes results. It is the insulin that makes them fat. It is the cortisol that elevates their blood sugar and sucks down their body protein content (catabolic effect).

Mainstream physicians are generally groomed into thinking that cortisol directly makes these owners fat. This is not true. First, the elevated cortisol directs the liver to begin creating and dumping blood sugar into the blood stream. This result is from the fact that when cortisol levels increase, the body thinks there is an emergency. When modern stress causes increased cortisol, no physical challenge ever comes and the body eventually recognizes the increased blood sugar as inappropriate. When the body recognizes the increase in blood sugar as inappropriate, insulin levels rise. When increased insulin secretes into the portal vein, because of body anatomy, the liver receives a higher message content to make fat. Increased fat synthesis leads to weight gain.

The second group of increased cortisol producers is often clinically missed because they lack body fat and this fools physicians. Any owner who has heart disease and appears on the emaciated side of physical body habitués deserves a deeper inquiry. Although they have increased cortisol output, they could have a 'wounded' pancreas. Wounded pancreases cannot increase insulin production to match the increased sugar output of high cortisol. Instead these owners rely on their IGF-1 to eventually bring their sugars back down to normal. Increased blood sugar proves only transient but the return to normal levels delays. The prolonged increase in blood sugar has oxidizing effects on the blood vessels (increased glycation rate, otherwise known as an increased rate of blood vessel rust formation). The normal fasting sugars are almost always normal, but the glucose challenge tests are not. These owners acquire blood vessel injury from the episodic elevations of blood sugar levels following the cortisol surges or carbohydrate binges. Increased blood vessel injury occurs from the rust processes that high sugar causes to the vessel walls. **These types of patients provide another situation for which insulin shots will prolong life because their insulin deficiency is at the root of their disease process.**

Further complicating these owner's health problems concerns the fact that they also crave sugar. They crave sugar because they lack sufficient insulin. Sufficient insulin is required so that the blood fuel can stay constant between meals. Normally, between meals the release of growth hormone occurs when blood fuel first begins to fall. The arrival of growth hormone at the liver that is stuffed full of glycogen, which insulin directed following the last meal, causes sugar, fat and IGF-1 release. The IGF-1 delivers the fuel to the hungry cells in the between meal state. However, in these types of afflicted owners their insulin release was inadequate, following their last meal, so their glycogen stores are insufficient. **Here lies the mechanism for another type of hypoglycemic prone owner.** These types present clinically as emaciated in their muscle mass and have very low body fat. They get hypoglycemic because they lack sufficient insulin to instruct the adequate storage of sugar within their livers to get them through the between meal state.

Thyroid Levels

Thyroid levels are a determinate of how fast the cell power plants can utilize blood fat. Owners with low thyroid levels have a diminished ability to utilize fuel in their power plants (mitochondria). Their livers are feeble because of the diminished thyroid message content at the level of the liver cell DNA programs. Scientific experiments on dogs that had their thyroids partially or totally destroyed clearly showed the pathologic changes at the level of the liver. Without adequate thyroid, the liver becomes pathologic because it lacks appropriate instruction at the level of its DNA programs.

The thyroid hormone also determines how many mineral pumps are present within the 100 trillion cell membranes. At rest, the largest amount of calories burn from recharging these 100 trillion cell batteries through activity of the mineral pumps. The thyroid message partially concerned itself with telling the body cells DNA program to manufacture more mineral pumps.

The more mineral pumps that exist, the more calories burned while an owner rests. Scientists call this the basic metabolic rate. The metabolic rate depends on two things: the amount of mineral pumps and the strength of the furnace flame within the mitochondria. Weight gain propagates until someone helps the thyroid hormone level to increase towards normal. Nuclear fallout, spent rocket fuel and other contaminants around the world have poisoned millions of owner's thyroid glands. Rather than educate physicians properly about the need to check the thyroid more completely, countless owners are falsely reassured that their thyroid hormone level is fine (see thyroid section).

Growth Hormone and the Unhealthy Liver

Unhealthy livers are unable to release adequate insulin-like growth factor (IGF-1) when growth hormone instructs them to do so. **The growth hormone can be diabetogenic in these situations.** Injured liver cells still make fuel available when directed to do so long after their ability to make proteins curtails. The liver manufactures the blood stream amount of insulin-like growth factor (IGF-1). There are many different reasons for the liver to become defective. Some of them include poor nutrition, excessive liver toxin exposure (alcohol, prescription drugs, putrefaction of the colon contents with liver reabsorption), poor trophic hormone levels (cortisol, thyroid, androgens, etc.) and excess iron storage. It is interesting that many medical textbooks emphasize that growth hormone is diabetogenic (promotes diabetes). Sensationalizing this rare possibility leads to yet another example of how physicians thinking gets groomed into the more profitable ways to treat disease. Instead of saying that growth hormone is diabetogenic when the liver is sick or the amount of growth hormone exceeds the ability of the liver to produce IGF-1 they propagate a half-truth. To tell the whole truth would facilitate a curiosity for the importance of IGF-1 for blood sugar control.

Insulin - Pro-Hormone Levels

Insulin is first made within a 'package', the pro-hormone, that converts to insulin before release into the blood stream. Certain situations create an increased release of the pro-hormone form of insulin (in the package). This occurs when the pancreas is forced to increase insulin production in response to some other body condition (obesity, high carbohydrate diets, insulin resistance). When this occurs the effectiveness to remove sugar from the blood stream diminishes per amount of pancreas secretion. The pro-hormone seems to retain some of its message content in the ability to make the liver create fat from sugar. This also happens in the growth of fat cells.

High Estrogen Levels Will Increase Insulin Need

There has been a flurry of media attention toward supposed new evidence about increased estrogen levels and the increased risk for heart disease. All along the evidence for this predictable association has been buried in the medical textbooks.

This chain of events is high estrogen leads to increased growth hormone release and simultaneously curtails the liver's release of insulin like-growth factor type 1 (IGF-1). The decrease of IGF-1 and the increased growth hormone message causes the sugar release to be uncompensated because there is a diminished release of IGF-1. In this situation, the blood sugar will rise higher than it does when excess estrogen is not present. The body eventually senses that the blood sugar is elevated and insulin secretes. The increased insulin secretion rate causes the liver to receive increased message content to manufacture sugar into fat and cholesterol. Fat and cholesterol secrete into the blood stream as LDL cholesterol. The blood has more LDL cholesterol and the triglyceride levels will be higher when measured. Triglyceride levels are only the summation of all the fat in the blood stream (the fat content of HDL, LDL, and VLDL cholesterol). Both high triglycerides and LDL cholesterol are associated with an increased risk of heart disease.

Low Androgen Levels

Many diabetic's disease traces back to a fall in their androgen levels for one reason or another. Early on in their disease if their physician recognizes this deficiency, when it occurs, total pancreas burn out can be avoided. In addition, many of the complications of diabetes such as: blood vessel pathology, colitis, and digestive tract dysfunction are made worse when androgen levels stay low.

Prediction: in the near future more attention as to whether insulin dependent diabetes resulted from liver injury first or pancreas injury first will come into the mainstream. The liver injury first variety has a component of causality in low androgens. Pancreas strain occurs whenever androgens fall because IGF-1 levels consequently fall off as well.

Physician's Sidebar

The important role of IGF-1 and the amount of GAG formation on the lining of the blood vessels, colon and stomach was previously mentioned. Briefly, IGF-1's presence in these tissues controls the important step of sulfation for these molecules creation. In fact the older literature called IGF-1, sulfation factor. A fall of the androgen message content in the body directly causes IGF-1 to fall off throughout the body.

A deficiency in this layer paves the way for an increased injury rate in these tissues. Add in the unmanaged high blood sugars of the uncontrolled diabetic and the injury rate accelerates.

Restoring the androgen level to normal would make sense as an additional step beyond blood sugar control alone.

Liver Facilitates the Assimilation of Fat from the Digestive Tube

There is a system in the liver for breaking down spent hemoglobin molecules by utilizing certain parts of these aged hemoglobin molecules, bilirubin, for the manufacture of bile. Bile facilitates the ability of the digestive tracts to absorb dietary fat and in the excretion of certain body wastes that the liver processes. The bile system is anatomically like a 'tree' in the liver. The smaller branches of bile containing the fluid coalesce into larger bile containing branches until they all converge in the common hepatic duct. The bile system tree has its root in the gall bladder that lies outside the liver. Salts of cholesterol are also used in the chemical concoction called bile (digestion chapter). All the constituents that make up bile facilitate the assimilation of dietary fat. When the liver becomes sick, the bile can back up into the blood stream. Jaundice results because of these components backing up in the blood stream. The hemoglobin breakdown product, bilirubin, being too high causes the skin color changes of jaundice.

The message content that the liver receives can help or hinder the fat content in the blood vessels and body. Equally important to overall body health concerns the molecular parts needed by the liver to deactivate numerous toxins.

The Liver Stores and Manufactures Vitamins, Minerals, and Hormones

The liver stores many vitamins such as vitamin D, A, and B12. In the case of vitamin A, the liver also manufactures its transport protein, retinol-binding protein. When carotene absorbs from the diet, it requires a capable liver to split carotene in half. Every carotene split in half creates two vitamins A's. Some owners begin to turn yellow when their liver becomes unable to split this apart.

Vitamin A is best thought of as a hormone. The vitamin A 'hormone' belongs to the most powerful class, the level one-type hormones. All level one hormones share the unique ability to regulate the DNA programs. Healthy owners have ample vitamin A stored in their liver.

Part of the ability to intake nutrition at the cellular level depends on the liver storing adequate potassium. A high potassium diet facilitates this ability. One potassium ion moves one sugar molecule out of the blood stream. Cells need adequate potassium to stabilize the proteins in the body cells. The size of muscle cells serves as an example of the importance of adequate potassium.

IGF-1 is a level two hormone made in the liver and its blood levels determine the ability of muscle and organ cells to uptake nutrients out of the blood stream. Once nutrients are inside a cell, they can be incorporated into power plant generation activities or regeneration projects. IGF-1 activity also depends on adequate potassium release from the liver. Normally potassium and IGF-1 release from the liver at the same time. Owners who possess high potassium and IGF-1 stores in their liver have a health advantage. The rate of IGF-1 manufacture in the liver is determined by the amount of DHEA and testosterone that reaches the liver DNA programs. The amount of IGF-1 released by the liver is determined by the amount of growth hormone released from the pituitary. **Common factors that encourage the release of growth hormone from the pituitary are exercise, fasting between meals, high pituitary dopamine levels, low pituitary serotonin levels and low blood sugar. Conversely, high estrogen depresses the ability of growth hormone to cause IGF release from the liver.**

Physician's Sidebar

Here lies the mechanism for how high estrogen levels cause insulin resistance (higher insulin need). Insulin resistance occurs because more insulin is needed when IGF-1 secretion rates diminish. This happens when estrogen, at high levels, inhibits IGF-1's release from the liver. The IGF-1 released diminishes despite the concurrent estrogen enhanced GH increase that directs the liver to dump sugar into the blood stream. This leads to the need for more insulin to return the blood sugar to normal. A diabetogenic effect of high estrogen operates in these cases. Instead, the textbooks propagate the half-truth of growth hormone being diabetogenic. While in fact, growth hormone only becomes diabetogenic when the liver is ill, high estrogen levels occur, or it secretes at abnormally high rates like in acromegaly.

IGF-1 essentially proves ineffective in the liver and fat cells for directing these types of cells in the taking of nutrition out of the blood stream. In contrast, insulin, is less effective for directing the uptake of sugar in cells like muscle, but many times more effective in directing the liver and fat cells to remove sugar for storage because of its lower blood stream concentrations and receptor concentration differences. Anatomically the liver always receives the highest amount of insulin message content whatever its secretion rate. A fat making consequence of increased insulin regards the fact that it is most effective in the liver and fat cells for taking nutrition out of the blood stream. This fact results from the liver and fat cells having the highest insulin receptor concentration of all cells in the body, around 200,000 per cell. In contrast, IGF-1 proves more effective for facilitating cells like organs and muscles in their up taking of nutrition. Increased estrogen then promotes the need for increased insulin to make up the deficit of IGF-1 in the peripheral cells. However, because increased insulin always hits the liver first the cholesterol and fat making machinery abnormally activate in these situations. This fact explains the increased LDL cholesterol and triglycerides seen in women on high estrogen prescriptions.

The Liver Manufactures Most Transport Proteins in the Serum of the Blood Stream

The liver makes most of the many proteins circulating in the blood other than the immune proteins. Many liver manufactured proteins are needed for diverse reasons. The liver manufactured proteins include transport proteins for the different steroids and minerals (iron and calcium). The liver also makes the clotting protein that prevents bleeding when a blood vessel tears. Lastly, the liver makes stress survival type proteins.

Common examples of liver manufactured proteins by type

1. Transport proteins

	molecules transported
Sex hormone binding globulin	estrogen and testosterone
Albumin	Some calcium and some progesterone
Thyroid binding globulin	thyroid hormones
Retinol binding protein	vitamin A
Cortisol binding protein	cortisol
Insulin-like growth factor **binding proteins**	insulin like growth factor
Transferrin	iron
Ceruloplasmin	copper

2. Blood clotting proteins

Factors 1-12

3.Acute phase reactant proteins that secrete during stress

Complement proteins
Fibrinogen
C reactive protein
Interferon
Ferritin
Ceruloplasmin

Chapter 12

Obesity-Related Blood Vessel Disease Resulting from Nutritional Deficiency

Often times, blood vessel disease is the direct result of varying types of nutritional deficiency. Most blood vessels will heal when nutritional deficiencies correct. The deficiencies injure blood vessels through rust promotion, hardening processes, or opportunistic mechanisms. Usually all three mechanisms are set into motion by nutritional imbalance.

Increased Blood Fat from Nutritional Deficiency

It has been asserted in the popular media that niacin supplementation may lower blood cholesterol. A more accurate statement would be to say that niacin is one of five nutritional cofactors that need to be present for optimal blood fat to occur. When all five of the cofactors are present, the body can absorb blood fat and burn it in the cellular power plants. When fat processes in this way, it creates the energy packets (acetate) that are needed for cell power plant function. This combustion process of acetate ends with the release of carbon dioxide and water.

When the other four cofactors are not present, many owners are condemned to failed attempts at natural healing. These owners often return to symptom control medicine ("the complex"). The necessity of these factors is well documented in the medical biochemistry textbooks, but the discussion occurs in a convoluted fashion. Knowledge of these fundamental 'must have' nutrients provides another way to heal. Pantothenic acid, carnitine, riboflavin, and Co enzyme Q10 are required for fat combustion in addition to niacin.

Without these factors, there is the tendency for blood fat to rise (LDL cholesterol) despite efforts to optimize insulin, cortisol, epinephrine, thyroid, estrogen, IGF-1, and androgens. Owners who eat a low carbohydrate diet, exercise aerobically, and participate in stress reduction measures may not improve their health if one or more of these cofactors is absent. These motivated owners fail to achieve their desired blood fat and weight loss because no one has counseled them on these basic nutritional cornerstones.

Living in these nutritional 'traps' becomes analogous to the creation of a smaller 'drain' for fat to exit once it has entered their blood stream. The drain enlarges when the cells have the nutritional mechanisms to combust it (see below). Optimal combustion of fat cannot occur without all of the necessary cofactors. When fat fails to combust, it builds within a cell and eventually spills backward back into the blood stream. Increased fat is analogous to a drain that can no longer dispose of its contents. One of the contents in the blood stream is fat. The inclusion of these factors in the diet or in some cases receiving them intravenously will allow a 'bigger drain' to form. The drain in this analogy involves the increased rate of fat removal made possible when the cells have the nutritional ability to process fat into carbon dioxide and water.

Owners who lack these five nutrients cannot properly access fat for the production of energy packets. This causes their cells to have power plant (mitochondria) problems because fat is the preferred fuel for many cell types. In fact, all raw fuel (protein, fat and carbohydrate) must first be converted to the fatty acid, acetate, in order for it to burn within the cell power plants. Because fat is the easiest to convert to acetate it constitutes the preferred power plant fuel of most body cells. The brain and red blood cell provide notable exception to this preference because they can only burn sugar for fuel.

Many owners do not receive sufficient amounts of these nutritional cofactors from their food. Multi vitamin pills may not solve this problem. When food is cooked and processed the five nutritional cofactors are destroyed. Absorption of nutrients constitutes a critical factor in the health of an owner. Disease processes and medications affect what absorbs and remains in the body. Empowerment comes from understanding why these factors play such a crucial role in the interrelationship between cellular power plants and energy extraction.

Carnitine is made from the amino acid, lysine. Vitamin C, vitamin B6, and SAMe (S-adenosyl methionine) are all needed for its manufacture. Note that SAMe is critical to its manufacture and becomes rapidly depleted without adequate folate, vitamin B6, vitamin B12, serine and methionine, the methyl donor system (see appendix). For each carnitine manufactured, three SAMe molecules are used. Meat contains various levels of carnitine. **Severe carnitine deficiency shows up as fatty liver disease and one form of heart failure.**

Carnitine, as a carrier molecule, delivers fat to the cellular power plant furnace for combustion. The liver, kidney, and heart cells need tremendous amounts of energy to perform. The cells cannot use the potential energy contained in the fat molecules without transportation by carnitine. Endurance athletes supplement carnitine for increased performance.

Niacin (vitamin B3) is the second factor necessary to burn fat in the creation of energy packets. Two other cofactors, riboflavin (vitamin B2) and pantothenic acid (vitamin B5) need to be considered with niacin. These three cofactors must be present in optimal amounts just inside the outer furnace (mitochondria). When all three are present the combustion of fat traps energy packets [ATP]. When any one of the three diminishes, there arises a progressive disability to trap energy. Heat is created instead of energy packets.

Pantothenic acid deficiency diminishes the rate of combustion of fat in the power plant. Fat can only be utilized when it is broken down two carbons at a time and attached to a pantothenic acid containing molecular machine, Co Enzyme A. This process cannot occur until carnitine delivers fat to the outside compartment of the cell furnace the mitochondrion. This acid forms the essential ingredient for the manufacture of coenzyme A. Coenzyme A is the carrier molecule that allows the orderly combusting of fat energy, two carbons at a time (as acetate). A deficiency in either carnitine or the above cofactors disables the refining process to acetate. When fat is the raw material for fuel combustion, these cofactors are required or body fat will accumulate in the blood stream. Pantothenic acid deficiency leads to a marked slow down in one's ability to burn fat calories.

The type of fuel that the mitochondria accept for combustion is restricted to one processed fuel type only, acetate. Certain nutritional cofactors are necessary in order to make acetate from the raw fuels of protein, carbohydrate, and fat. Many owners have power plant problems because no one counsels them about what these cofactors entail. The refined fuel requirement is similar to power plants of the physical world. These structures can only combust one specific fuel type for operational purposes (natural gas, coal, radio active material, or fuel oil, etc.). Whether raw fuel starts out as protein, fat or sugar, it needs to be processed into acetate (the simplest fatty acid) before it can be combusted in the mitochondria. Acetate is the only fuel that the body power plants can utilize aerobically. When acetate burns in the presence of oxygen, carbon dioxide gas, water and energy packets are created. The cell power plants only accept acetate for burning within the mitochondria.

Coenzyme Q10 (ubiquinone) is the last nutritional Cofactor. It is similar to niacin, riboflavin, and pantothenic acid, but Coenzyme Q10 allows for additional trapping of more energy [ATP] within the cell. **Co enzyme Q10 is special in that it is made from the same enzymatic machinery as cholesterol.**

Without adequate Co enzyme Q10, the cells are compromised in their ability to generate energy packets and make more heat energy. This energy is waste. Without Co enzyme Q10, energy packet (ATP) formation diminishes.

Extreme energetic compromise occurs when the heart is without sufficient coenzyme Q10 because it constitutes a vital component, within each of its cell's 2000 mitochondrion. It is necessary to effectively trap energy within the power plant.

Each heart cell contains around 2,000 mitochondria. This many mitochondria make sense since the heart's pumping action requires tremendous amounts of energy to sustain the force of each beat. Adequate Co enzyme Q10 is a determinant of the performance of a heart cell. Lowered work ability leads to lower cardiac function. The heart has the highest needs for coenzyme Q10 in the body. Huge amounts of energy packets are needed here. Each heart cell has thousands of power plants (mitochondria) that need all five cofactors to maintain healthy cell function.

Popular cholesterol lowering drugs known as the statin class inhibit Co enzyme Q10's production. The body's sufficient attainment for this nutrient is further compromised because coenzyme Q10 is unstable within most processing techniques for food.

There is suspicion that people on these drugs tend to die at about the same rate as the untreated groups, but for different reasons. This is more alarming when adding in the suspected increase in cancer rates for owners who take these drugs.

The explanation could be from the fact that initially, cancer cells have an inferior ability to generate energy compared to healthy cells. When the body is healthy this is a major advantage within the immune system for destroying cancer cells at an early stage. Most immune systems continuously destroy cancer cells by generating high energy burst and targeting these packets at cancer cells. This exposes a link for why those taking statin drugs could be at increased risk for cancer. These drugs tend to deplete the body content of coenzyme Q10 and hence, cells like immune system cell types have diminished ability to out perform the cancer cell in energy generation.

This does not mean that statin drugs are never indicted for the prevention of heart disease. Many owners insist on taking a passive role in their disease process. For this group, their doctor has very few alternatives. Those owners willing to take responsibility for their health can use therapies without these side effects. These remedies can often times avoid statin drug usage and its potential side effects. The inclusion of high quality coenzyme Q10 would be a big help in the clinical situations where the statin drugs are needed. This need does not often occur in a motivated patient.

These factors need to be present nutritionally in order for fat to be utilized in the production of energy work packets. The utilization of sugar as fuel has other requirements.

Accessing Carbohydrate Energy for Cellular Needs

Carbohydrates can only be burned anaerobically (without oxygen) when any of five nutritional cofactors are deficient. This leads to massive increases in lactic acid production and fatigue. Fatigued cells have difficulty defending from rust and hardening processes. In the blood vessel lining cells this leads to an increase in Velcro formation.

When the five nutritional factors are deficient, the cell has no way of obtaining the necessary sugar fragment, acetate. This fragment can only be combusted within the mitochondria in the presence oxygen.

In order to burn sugar (carbohydrate) to carbon dioxide and water, the cell needs five types of nutritional molecules in sufficient amounts plus adequate oxygen in the combustion chamber. Lipoic acid, riboflavin, niacin, pantothenic acid, and thiamine, in the presence of adequate magnesium constitute these factors.

If the first part of sugar breakdown occurs without oxygen, a three-carbon fragment, pyruvate, is formed. Only when there is adequate oxygen and the five above factors can this molecule enter the power plant for further energy release. With a deficiency, pyruvate breaks down to lactic acid. The liver usually clears this acid from the blood stream, but when formation becomes excessive, it builds up in the muscles to prevent death. The enzyme, pyruvic dehydrogenase and these five factors cut the head off (make a two carbon molecule, acetate) of pyruvate before it can form lactic acid and stabilize on the carrier molecule Co enzyme A (formed from pantothenic acid). The larger share of energy contained in a sugar molecule cannot be utilized unless all of the factors are present.

There are many owners in pain and experiencing chronic fatigue only because their cells have a decreased ability to burn sugar energy in a healthy way. The smallest physical exertion condemns these owners to bed rest. Their tissues are full of lactic acid and this creates pain. These patients need to find a competent nutritionally oriented physician or they will continue to suffer. Not all chronic fatigue and muscle aches are from this cause, but a significant percentage is due to nutritional deficiency.

Amino acids can be used as fuel only after the liver converts them into sugar in a process called gluconeogenesis. All amino acids contain at least one nitrogen group (the amide). This group must be removed and eventually converted into urea in the liver for excretion by the kidneys. Adequate urea helps the kidneys concentrate their waste and conserve water. Cortisol, epinephrine, and glucagon hormones encourage gluconeogenesis. Epinephrine and glucagon have little effect without adequate cortisol. Insulin and growth hormone oppose gluconeogenesis. Amino acids are another source of raw fuel that needs refining to acetate before becoming combustible.

The Basic Unit of the Human Form is the Cell (optional reading)

The numerous body cell types are analogous in some ways to the many different varieties of fruits and vegetables. The different cell types are composed of different textures of skin, color, shapes, sizes and consistency.

Extending this analogy one can envision the different skin types of these foods as analogous to the different types of environments that surround the cell types. Cartilage and bone cells that are widely spaced between the hard matrixes that they create are like cocoanuts. Red blood cells are like sponges in consistency (technically not a edible food, but the consistency and resiliency are accurate.).

Some owners have health problems because their cell types are like overly ripe fruit or vegetables. Aged fruits and vegetables have deteriorating qualities. The cell types also have deteriorating qualities that are observed in humans as the path to old age. The goal is to keep all the body fruits and vegetables in their most appealing state (discussed in book three).

Almost all have a complete computer program (DNA), but only certain programs are active within a cell type. The programs that are active determine what a cell looks like and the products that the different cell factories produce. The activity of the DNA program (genes) also determines the types and amounts of the different cellular machines (enzymes) that perform different assembly line activities. The cellular factories (Golgi apparatus, endoplasmic reticulum, etc.) need power to generate their activities. This energy is supplied by the cellular power plant (mitochondria). When a cell is healthy the energy packets (ATP) created by the combustion of fuel and oxygen are trapped in the cell. Many cells have additional waste incinerator facilities (peroxisomes). All cells have a cellular force field (plasma membrane). The cell membrane supplies energy for work, derived from available ATP, and shields the cell from harmful, foreign molecules. The supporting architecture for the cell's interior is called the cytoskeleton. The supporting architecture for the cell's exterior is called the extra-cellular matrix. When the supporting architecture diminishes the body assumes the flattened look of old age.

Chapter 13

The Fear of Real Hormones and the Reassurance of Fake Hormones

Millions of women take altered message content in the form of their hormone replacement therapy. Of course their doctors and they do not realize that their prescriptions contain altered message content delivered to their trillions of cells. The wrong message instructing the DNA programs leads to side effects and toxicities. Basic science says this is true. However, the discussion of these facts within the textbooks occurs in a disjointed and fragmented fashion. Textbooks written in this way keep doctors in the dark about what science really knows about hormones.

Female hormone replacement science is discussed in more detail in *The Body Heals.* Here the focus concerns the advantage of real hormone replacement over fake hormone replacement. Real hormones carry accurate message content. Fake hormones carry inaccurate message content to the cells. The altered message results in side effects and toxicities. Real hormones cannot be patented and hence cannot be as profitable. Hormones become fake when the real hormones shape becomes changed in order to obtain a patent. No shape change then no patent becomes possible. The precise molecular shape contains the hormone message content. Simple science discussed here.

The miracle of convincing doctors that fake hormones are better than real hormones doesn't mean that the side effects are not real. Most doctors have been educated that fake hormones are best.

About one in nine women will develop breast cancer. Probably a significant part of this occurrence results from fake hormone usage and excessive exposure to hormone mimics in the environment. Yet, if a woman gets cancer while on fake hormones it is not considered malpractice. In fact many doctors get away with prescribing fake hormones without a thorough laboratory inquiry of a patients hormone status. However, the physician that takes the time to perform a thorough laboratory inquiry into a patient's hormone status and then prescribes real hormones considers himself/herself at risk for a malpractice verdict if that patient gets cancer.

There are several possible explanations for why the standard of medical care, in regards to hormone replacement therapy, has become so ludicrous.

1. Physicians do not receive proper education about the fact that the precise shape of a hormone carries its message content. A change in the shape, in order to get a patent, changes the message content. Simple science but not commonly taught in a way that doctors can see the problem of fake hormones.

2. The public lacks awareness that holistically trained physicians exist who can prescribe real hormone replacement therapies. The real hormone replacement therapies need to be filled by a compounding pharmacist. Compounding pharmacist specialize in real hormones.

3. The media and the medical industrial complex have a cozy arrangement. Each benefits. One from increased advertising revenue and the other form increased drug or

procedure sales. One of the most effective media strategies creates fear about real hormones in both the doctors and patients minds. However, remembering #1 above helps to see that fake hormones cannot be better than real hormones because of simple science. Only real hormones can deliver accurate message content to one's trillions of body cells.

One crucial step for the turning off of the fat maker machine concerns the seven hormone types, which allow or curtail fat accumulation. As long as fear propagates about real hormone replacement therapies a powerful weight loss tool will go underutilized in those that have severe glandular deficiencies.

Almost all cases of obesity have a large glandular component. Obese people are in general not lazy, stupid or unmotivated. Obese people have at their core a defect in one or more of their glandular secretions. The wrong hormones direct inappropriate body energy into storage as fat (see text).

Through proper nutrition, exercise training, and stress reduction measures many overweight owners heal. Some of the weight loss effect occurs because these measures improve the glandular secretions. But how does the doctor rectify those owners whose glands have big defects? Despite their best efforts for a weight loss effect they fail. It becomes a personal choice about deciding to go ahead with real hormone replacement therapy.

Cancer rates increase with each passing decade. Most of this increase probably stems from nutritional deficiencies, environmental toxins loads, and immune dysfunction. Yet, real hormone replacement takes the blame wherever possible.

The renaming of somatotropin to growth hormone helps to instill additional fear. Also the renaming of the nonsuppressible insulin like activity of the blood to insulin like growth factor type one (IGF-1) helps also to instill more fear still about growing cancer, as well.

Never mind the fact that the highest insulin receptor concentration occurs on the liver and fat cells at about 200,000 receptors per cell. Breast tissue composes itself mostly from fat.

Overweight females are generally considered to be at increased risk for developing breast cancer. High insulin levels are a must before an owner can gain body fat. Yet no one counsels these breast cancer victims about insulin reduction techniques. Instead IGF-1 gets the blame as a supposed risk factor for tumor growth.

If this were so then athletic training, adolescence, and pregnancy all would be risk factors for developing cancer. In all three conditions IGF-1 level are several times higher than middle-aged people. However, epidemiologic studies show that each one of these situations provides protection and does not seem to be a causative factor for cancer development. Also as IGF-1 decreases with age, the health decreases as well.

Some overweight owners have severe glandular defects with resulting decreases in growth hormone and/or androgen secretion. Either one of these defects will lead to decreased IGF-1. When IGF-1 decreases then insulin must increase (see text). The more insulin, the more the fat maker message within the body. A rising insulin level and a falling IGF-1 more often occurs when cancer develops rather than the reverse.

Before a prescription of either growth hormone or androgen hormones can be entertained, the liver function and adrenal function need to be evaluated and found to be normal. These types of hormone replacement will cause problems when either the liver or adrenal health are not attended to. Many doctors fail to counsel their patients that growth hormone naturally increases when blood sugar falls. In uncontrolled diabetes and chronic grazing type feeding habits the blood sugar never falls so growth hormone levels fall.

Growth hormone injections need to occur when the blood sugar is already on the low side. Like first thing in the morning or just before exercising provide such examples.

Insulin deficiency causes one form of diabetes.

IGF-1 deficiency causes one type of syndrome X type associated obesity.

Societal programming makes insulin injections seem like a no-brainer.

Yet, societal programming makes growth hormone and/or androgen replacement therapy, prescribed to raise IGF-1 levels, seem risky. Just like the insulin deficient diabetes patient, the IGF-1deficient caused syndrome X patient will die more quickly without treatment.

Think about the billions of dollars of lost revenue from the diminished need for prescriptions and procedures if a syndrome X patient received treatment.

Truncal obesity, high blood pressure, and type A personality should be followed with:

- Fasting insulin and C-peptide level
- Fasting IGF-1 level
- 24 hour urinary steroid and thyroid profile (including estrogens for men and women)
- Baseline PSA level for men over forty
- Hemoglobin A1C

Growth hormone and androgens are contraindicated for most cancers. The history of cancer proves a relative contraindication that your doctor will decide about on an individual basis.

If exercise, nutritional therapy (see text) and stress management fail to improve the above first three-baseline labs real hormone replacement therapy should be considered. The glandular secretions must improve before a weight loss occurs.

Chapter 14

The First Day of Your New Life

Life changing events always involve an increase in awareness. Where once one was unconscious, they are now conscious. Poor health largely results from many years of successive unconscious choices that accumulate into consequences. Healing always involves awareness. Awareness allows focus. Sustained focus leads to positive health results.

My own life-changing event involved the realization that I had a pretty good chance for developing heart disease. Both of my grandfathers had Syndrome X. Both of them died of massive and sudden heart attacks.

For years, I dutifully followed the latest mainstream medical diet advice about low fat and increased carbohydrate consumption. Predictably my cholesterol and triglycerides worsened. My belt size increased six inches.

Years ago, my patients provided me with the first clues that my education contained grave inconsistencies. They were the ones who insisted on trying low carbohydrates but high fat diets. Back then, I was shocked by their lab results that documented a fall in their cholesterol and triglycerides. In those days my education level could not explain this contradiction for why a high fat and protein but low carbohydrate diet leads to a fall in cholesterol and triglycerides. It has taken me years of study to identify the hormones that this diet promotes and why cholesterol and triglycerides levels decrease (see text for specifics).

My next big step of awareness involved my preference for sweets and other carbohydrates. Once I realized that high insulin levels that the high carbohydrates promote were a stimulus for fat cell growth within my arteries, a mental image arose. Each time I began to reach for my favorite carbohydrate, I pictured my arteries growing fat on their inside surfaces. My new awareness curtailed my life-long craving for these insulin producing foods. My cravings left because I knew that they would eventually kill me off.

Several years into my forties, I began to notice a consistent elevation in my blood pressure readings. Despite vigorous exercise it persisted. I began to ponder why a body would elevate its blood pressure. This line of thought led to connecting scientific facts within the mainstream textbooks as to why blood pressure elevates.

For me, the solution to my blood pressure problem involved correcting my mineral imbalance, decreasing my insulin needs, recharging my methyl donor system and increasing my nitric oxide production. In six months my blood pressure fell by over 30 points (explained in *The Body Heals*).

The most recent step in my awareness towards my own healing involved that glandular failure was possible. I also needed to realize that gland function centrally effects health. In many cases glandular failure responds to nutritional support, detoxification, lifestyle changes, and diet improvements (see text).

However, some overweight owners glands have severe defects. Obesity will not resolve in these cases until these defects are identified and corrected in as safe a way as possible. **The initial hormone report card includes:**
- Fasting insulin and C-peptide
- Fasting IGF-1

- Adrenal steroids (including aldosterone)
- Androgen type steroids made in the gonads
- Thyroid hormones' levels (including reverse T3)
- Estrogen status
- Homocysteine level
- Prolactin level
- PSA level for men over age 40
- Hemoglobin A1C

Note of caution to clinicians: labs routinely over the last few years have been arbitrarily lowering the 'normal' range for IGF-1 levels. Not so long ago the normal range for IGF-1 was 250-400 ng/ml. Currently some labs say that values as low as 80ng/ml are normal. This is ludicrous if one desires health and weight loss. The lower acceptable level for IGF-1 will probably end up being around 250ng/ml. All physically fit and youthful people have at least this value and many have much higher levels.

The medical legal rule about standard of care makes it difficult for conventionally trained M.D.'s to practice holistically in their replacing glandular deficiencies with real hormones.

Any treatment plan for a given ailment carries risk. This is true whether the plan is holistic or based in the mainstream paradigm of symptom control. Holistic approaches carry less risk in general because they seek to heal the root cause of a problem. In contrast, mainstream medicine's standard of care protocols usually involves increased risk because they treat the symptoms of disease and not its cause. Symptom control always has side effects and toxicities. Yet this is the legally sanctioned approach that mainstream-trained doctors are intimidated into following lest they find themselves outside the standard of care protocols. Never mind the facts contained in this and other books that suggest that the standard of care has less than perfect motives for what it touts.

As discussed previously, one in nine women will develop breast cancer. If before cancer develops they receive a superficial hormone evaluation and then are prescribed horse estrogens cycled with fake progesterone this is felt to be the standard of care. As long as the standard of care is upheld the risk to the doctor for a malpractice case is minimal. However, if these same women were to receive a thorough hormone evaluation and steroid deficiencies were identified and treated with real hormones, the onset of cancer would be sometimes considered due to medical malpractice.

Even though basic science shows that real hormones to be safer than fake hormones the medical legal standard supports the later. It becomes a personal choice as to whether one with documented severe glandular defects decides on taking real hormones replacement therapy.

I feel the best course involves informed consent. Informed consent for hormone replacement therapy involves reading this book and others like: *Grow Young with HGH, Some Things Your Doctor may not tell You about Menopause, Natural Hormone Balance for Women, Safe Uses of Cortisol, and Hormone Replacement.* The authors and the publishers of these fine books are in the back of this book.

Informed consent also involves a balanced discussion about the medical industrial establishment's negative feelings about real hormone replacement therapies for middle-aged diseases like obesity. Patients need to be informed that mainstream medicine frowns upon growth hormone and testosterone replacement therapy for overweight owners. They purport to have ample unbiased studies that link these therapies to cancer's development and accelerated growth.

Yet, they fail to mention or consider several inconsistent details. If all the negative fear about growth hormone were true, a good nights sleep, youthfulness, fasting, and athletic training would all be an increased risk factor for developing cancer. All of the above are well documented to increase growth hormone release and hence IGF-1 as well. However, epidemiological studies have shown these all lower the risk for cancer's development.

Each overweight owner needs to carefully consider the official view (standard of care) with the real hormone replacement viewpoint. An informed consent form needs to be signed, before beginning treatment, acknowledging the acceptance of risk for the development of cancer according to the mainstream viewpoint. Also, an agreement about keeping their doctor up to date about their condition needs to be worked out. In addition, regular follow up needs to occur.

Recognize that mainstream physicians habitually discredit that which they do not understand. In these cases show them this book or the others listed above. Kind and informed patients can help their doctors to learn a new.

Reading self-help books like this one can bring about empowerment for what health care choices are available. With empowerment comes personal responsibility for what is still unknown. Health care choices currently have many unknowns.

The practice of medicine involves probabilities. I feel that it is more probable that real hormone replacement therapy for documented defects is safer than ignoring the defect, prescribing fake hormones or engaging in symptom control .I base my belief on the contents of this book and others listed in the bibliography. I also want to acknowledge the current state of the unknown and the consequent incomplete healing choices.

In the end, each reader prays about what the right course is for him or her to take.

My future writing will undoubtedly contain new caveats, insights and cautions. Such is the nature of this life as we progress into the light.

Good luck and God bless.

Appendices

Appendix A

Hierarchy of Hormones - Informational Substances

Before proceeding with this book, it helps to have a concise grouping of how the different hormones can be separated into a successive hierarchy. The hierarchy between the different hormones concerns their degree of influence and length of action when compared to one another. There are four basic groups of successive levels of influence on how the body spends energy. Hormones direct cell energy usage. The more powerful the hormone group, the more central it is in the energy direction where it arrives.

It helps to elucidate the common error of how mainstream medicine often focuses on the 'lesser' hormones. The lesser hormones get the press coverage while the more powerful hormones are only peripherally addressed. By taking this approach to chronic disease treatment strategies, symptom control is all that is possible. Symptom control medical approaches always produce side effects and have nothing to do with healing. By adding back the hormone hierarchy these errors become easy to expose.

Grouping the different informational substances (hormones) into four groups will help in understanding the interdependent relationship of the many various hormones. Also, many chronic degenerative diseases will begin to have healing solutions.

Common degenerative diseases amenable to healing once the hormone hierarchy is addressed are: adrenal gland caused asthma, rheumatoid arthritis, systemic lupus erythromatosus, osteoarthritis, adult onset diabetes, obesity, some cases of heart disease, ulcerative colitis, Crohn disease, some liver diseases, some kidney diseases, some cases of senility, and muscle wasting diseases.

All of these diseases have an increased likelihood

of a solution when the clinician attends to these disease treatments in a logical progression from the most powerful hormone imbalances down to the weakest. When supplementation strategies are needed, real hormones are used because only they contain accurate message content. The body was designed for real hormone message content. Whenever the altered shaped of the hormone substitutes are given in place of real hormones, there are always side effects. When the shape changes, to get a patent, the message content changes. Simple science discussed here but not commonly taught in a way that physicians can understand.

The four groups in the hierarchy of hormones are:

Group 1 - the supreme commanders of body energy

This powerful group of hormones directly switches off and on different DNA programs. The DNA programs activated in a cell

determine the activity of the cell and the degree of repair (rejuvenation). The activity and state of repair determine the usefulness of the cell. When cells are given good informational direction from the level one hormone class, they will be productive and healthy. The thyroid hormones, steroids, and vitamin A comprise the level one hormone class. The quality of the mixture and amount of level one hormones reaching the over one hundred trillion cells forms a central determinant of health. An owner's health is powerfully influenced because these hormones all contain message content that instructs the DNA (genes) program activity.

The level one hormone class (steroids, vitamin A and thyroid hormones) comprises the only hormones powerful enough to directly interact and instruct cellular genetic material (the DNA). Quality of the proportional mixture between these hormones determines the highest level of energy expenditure. At this primary level these hormones quality of presence determines whether a cell uses available energy efficiently or not. **Healthy people always have sufficient rejuvenation message content to counteract the ongoing injuries of life. Unhealthy people do not. Appropriate level one message content allows health to continue.**

`This class of hormone has access to every body chamber. All healthy owners predictably have high quality message content amongst the level one hormone class. Conversely, unhealthy owners predictably have lousy hormone quality as part of their aging process. These owners will continue to age prematurely, until someone helps these owners regain more optimal message content.

Group 2 – the amino acid chain

All hormones message content is about directing cellular expendable energy, likewise with the level two hormones. Level two hormones deliver a message when they bind to a cell surface receptor (their own unique type) or they bind a receptor inside the target cell. The level two hormones are only able to influence cell energy expenditure on existing cell structures and enzymatic machinery activity. They are not able to direct DNA programs in the manufacture of new cell structures or new enzymatic machinery. Only level one hormones are powerful enough to do that

Insulin and glucagon provide examples of level two hormones. Like other level two hormones, they are made up of a specific sequence of amino acids that are twisted around in three-dimensional space. The shape of these specific sequences and resulting twist contains precise message content. Insulin and glucagon are also an example of level two hormones that contain opposite message content. The message content difference between insulin and glucagon directs the target cell to spend energy in the opposite way. Consistent with level two hormones, each can only affect existing cellular enzymatic machines or structural content. Insulin turns off the enzyme machines that glucagon turns on. Likewise, the message content of glucagon turns off what insulin turns on. Each of these opposing hormones has enzyme machines that they activate. The direction of energy within a cell determines which enzymes are quiet or active. Hormones direct which way energy moves in a cell. Catabolism versus anabolism describes an example of opposite energy movement.

The liver cells prove instructive as an example for the opposite energy direction between the message content of insulin and glucagon. Insulin directs liver cell energy into fuel storage (anabolism). Fuel is stored in various locations, but insulin stimulates the liver in the manufacture of glycogen, triglycerides, cholesterol, and LDL cholesterol. Insulin also inhibits the liver from turning amino acids into sugar and fat. The more insulin in the body, the more these activities occur within the liver. Conversely, glucagon directs the liver to release stored fuel and curtails new cholesterol and triglyceride manufacture (catabolic effect). It also stimulates the liver contained enzymatic machines to make sugar from available amino acids.

Some level two hormones are further endowed with the ability to leave a 'last will and testament' before being chewed up into its component parts by intracellular machinery. Intracellular machinery eventually dismantles the level two hormones. There is an intermediate step. The last will and directive of some level two hormones occurs between activating their receptor and being dismantled. The last will and directives are additional messages to the cell created by level two hormones activating level four hormones. These messages exist for a limited time. During that time, the level two hormones affect which level four hormone precursors a cell will receive message content from (Sears).

Insulin and glucagon provide good examples of how a hormone's presence determines which level four hormone precursors are formed. Also consistent is the fact that level four hormones precursors, stimulated by insulin, have the opposite effect on cell energy, when mature, as the direction that glucagon level four hormones impart (see group 4 below).

Group 3 – the subservient hormones

The level three hormones are composed of single amino acids, which have been structurally modified or are from short chains of amino acids. These hormones depend on directives of the higher hormones 'stage setting'. The level one and two hormones set the stage for structural integrity and enzyme machinery activity contained in a cell. The stage setting is centrally determined by the level one hormones at the DNA level.

In turn, the level two hormones determine the activity of the setting, the enzymes. The level three hormones channel blood and fuel to the cells. However, the level three hormones depend on level one hormones directing the manufacture of their receptors. Without enough level three-hormone receptors being made, many chronic diseases begin (asthma from epinephrine receptor deficiency). The level four hormones can only affect the 'props' within the cell (see below).

Biogenic amines, cytokines, and endothelin all belong to this class of hormones. Dietary deficiencies can effect development and manufacture of these hormones. Many different vitamins prove important for biogenic amine's manufacture. In general, the level three hormones have an effect on the caliber of blood vessels and properties of the cells in these vessels in a certain area of the body.

The level three hormones (muscle chapter) delivering message content in a certain blood vessel determines much about the oxygen availability, nutrient delivery, waste removal, stickiness of the vessel wall, clotting tendency, and immune cell behavior.

Group 4 – local acting, subservient hormones

These hormones exist for seconds, only long enough to deliver message content to nearby cells. Within seconds of their release they are deactivated. These hormones only having an influence for both the duration and as far as the sound of their 'voice' will travel. This is in contrast to the other three levels of hormones that can travel more extensively in delivering their message content.

Some of the level four hormones, like nitric oxide gas, have powerful penetration abilities through body barriers. Although nitric oxide has powerful penetrating abilities, its message content delivery ability is limited due to its rapid deactivation. This short lifespan diminishes the distance of influence of nitric oxide's message content.

Other level four hormones, like the ecosanoids (hormonal fats) that include prostaglandins, leukotrienes, and lipoxin are also limited by their short life spans. The ecosanoids come from essential fatty acids (hormonal fat precursors). These include Linoleic, linolenic, and arachidonic acid. All three of these essential fatty acids can only be obtained in the diet. The type of diet determines which level two hormones message content will predominate.

An example of the opposing possibilities of level two hormones and how they determine what level four hormones are possible occurs between insulin and glucagon. The opposite message content contained between glucagon and insulin has a powerful influence on which level four precursor hormones are possible. High carbohydrate diets tend to promote the pro-inflammatory hormonal fat precursors to line many cells. Certain low carbohydrate and high

269

protein/fat diets promote anti-inflammatory hormonal fat precursors (see *The Body Heals*). The type of level two hormones limits the possibilities for which level four hormone precursors line numerous cells. Those hormonal fat precursors that line the surface of cells determine the likelihood of developing or preventing some malicious chronic degenerative diseases. Some diseases effected by the presence or the absences of hormonal fat precursors are cancer, heart disease, arthritis, depression, fatigue and immunodeficiency syndromes.

There are three principles about the ecosanoids (hormonal fats) that are important. First, only green plants can make two out of three of these hormonal fat precursors called the essential fatty acids. Animals like salmon that live in the wild eat plankton and therefore contain significant amounts of these two plant manufactured essential fatty acids. Farmed, grain-fed animals are raised without these essential fatty acids

presence. The green plant manufactured essential fatty acids are linolenic and linoleic acid. Free-range chicken eggs are good sources for these plants derived essential fatty acids. Wild game such as elk and deer are also sources for these two essential fatty acids. These are good sources for essential fatty acids because of the high content of green plants in their diet. Green plants are the source of the anti-inflammatory fatty acids, linolenic and linoleic acids. Non-green roots and grains are more likely to contain the pro-inflammatory arachidonic acid. Store bought chicken, meat and eggs, because the animals are grain fed, will be higher in the arachidonic acid type.

Second, eating all the right essential fatty acids can still output the wrong hormonal fats. These fats are subservient to the more powerful higher hormones. Level two hormones like insulin and glucagon determine which hormone fat precursors become created from essential fatty acids in the diet. The essential fatty acids in the diet are the raw material created for precursor hormonal fats. The balance of message content between insulin and glucagon predetermine which precursors are created.

Third, hormonal fat precursors are ideally contained on all the cell surfaces. When released, these hormones have a limited area of influence because of their short lifespan. Scientist denote this shortened hormone lifespan by the term paracrine. Paracrine hormones can only deliver message content in their immediate area of release. All level four hormones are paracrine in nature. The level four hormones are only released when triggered by higher hormones. Sub-optimal hormonal fat message content has a powerful influence on several chronic diseases.

The hormonal fats are so important that many powerful medications work by poisoning their ability to be produced. Some medications that work by poisoning hormonal fats are aspirin (and other nonsteroidal anti-inflammatories) and related non-steroidal anti-inflammatory medications, cortisone derivatives, and the newer cox 2 inhibitors. These symptom control approaches have consequences to the balance of fine tuning abilities occurring in cells. All of these redirect body energy. With an understanding of hormonal fats and how to optimize their function, one can work with their body and begin healing.

Nitric oxide has a powerful ability to lower blood pressure and is also within this fourth class of hormones. It can only briefly increase blood delivery into its local area. This is how the medication Viagra works and why all the warnings about it lowering blood pressure are given. The medicine content of nitric oxide increases blood flow to the penis. This natural blood flow regulator requires the presents of the nutritional co-factors arginine, riboflavin, tetrahydrobiopterin, and thiols for its biosynthesis. Thiols are contained in garlic and onions. Tetrahydrobiopterin derives from folate. How many hypertensive and impotent owners originate simply from these nutritional deficiencies?

Be aware of the hierarchy of different hormone groups and what dietary factors are required for them to be produced. Owners who have optimal balance at each level are in good health. Conversely, a lowered quality of informational content occurring at any of the successive levels or dietary deficiencies leads to health consequences. The science exists to help owners receive better hormones and good nutrition.

Appendix B

The All Important Methyl Donor System

Specific deficiencies in the adrenal gland can initiate the hypertension disease process. The owner using nutritional strategies can successfully treat these deficiencies. These nutritional strategies can reduce the owner's blood pressure and lower the owner's heart disease risk profiles, as well. This first adrenal deficiency arises out of **the methyl donor deficiency syndrome**. Deficiency caused diseases of the various components of the important methyl donor system includes both high blood pressure and heart disease states.

Differences in the message content between epinephrine and norepinephrine (both are known as catecholamines) exist. Alternatively these are called adrenaline and noradrenaline, respectively. When certain nutritional deficiencies develop epinephrine decreases first and concurrently more nor-epinephrine production occurs within the adrenal gland. This situation leads to significant physiological consequences within the blood vessel. High blood pressure provides one example for the differences in message content between norepinephrine and epinephrine (see below). The altered blood vessel performance then increases the risk for heart disease from the resulting high blood pressure.

The adrenal gland divides into two components: the cortex and medulla. Both parts of the adrenal gland secretions message content concern the survival of stress. The better the secretion function within both adrenal components, the better an owner can survive the stresses of life without physiologic consequences. High blood pressure provides one example of the physiologic consequences when the adrenal secretion quality diminishes.

Biological, emotional, and environmental stresses are hard on both compartments of one's

273

adrenal gland. Stress constantly triggers the steroid producing outer section of the adrenal gland. This causes cortisol to be secreted (among other steroids). Only when sufficient cortisol circulates in the blood stream can the cell receptors recognize epinephrine and nor-epinephrine message content. Epinephrine and nor-epinephrine secrete from the adrenal medulla with stressful stimuli. The dependence of the adrenal medulla for adequate cortisol from the adrenal cortex proves important. Without cortisol preparing the manufacture of the receptors for these adrenal medulla hormones [epinephrine and nor-epinephrine] their message content will go unrecognized (explained in *The Body Heals*). This subsection will only address the nutritionally induced imbalance between epinephrine and nor-epinephrine secreted by the adrenal medulla.

Normally the adrenal medulla secretes 90% epinephrine and only 10%

nor-epinephrine. Epinephrine is preferred because it opens up the blood supply to the heart, skeletal muscles, and the liver. In contrast, nor-epinephrine does not do this. All other blood vessels (except the brain where blood flow is kept constant in the healthful state) are directed to clamp down when nor-epinephrine provides the message (hormones contain message content). In contrast, the net effect of epinephrine message content (because the muscle and liver blood vessels percentage of total body blood vessels are so large) results in a lowered peripheral vascular resistance. This effect leads to a lower diastolic blood pressure. Epinephrine has this effect within the body even when the epinephrine level within the blood stream reaches relatively high levels. Epinephrine also increases cardiac performance, which may result in a slight rise in systolic blood pressure.

The effects of epinephrine within the heart, liver and skeletal muscle vessels directly contrast with the message that nor-epinephrine delivers to these same blood vessels. Nor-epinephrine's message content directs the vessels to clamp down on everything except the blood supply to the brain. When blood vessels constrict without a corresponding dilation somewhere else blood pressure elevates. Healthy owner's adrenal medulla's (inner adrenal layer) manufacture and secrete 90% epinephrine and only 10% nor-epinephrine. However, when the adrenal medulla experiences deficiencies in certain vitamins and cofactors this ratio changes for the worse. A nutritionally caused inability to make epinephrine will tend to raise blood pressure when nor-epinephrine secretes instead of epinephrine.

Both of these hormones act rapidly and effectively for the redistribution of blood flow. This effect causes the blood stream to maintain an adequate flow of blood within the brain during normal body movement. Sufficient epinephrine or nor-epinephrine needs to secrete or unconsciousness results. For example, when an owner gets out of bed the forces of gravity cause the blood pressure to suddenly drop. Healthy owner's adrenal medullas make epinephrine in sufficient quantities to deliver a smooth machine that goes from lying flat to standing upright with grace and ease. Normally these hormones have a lifespan of about 2 minutes within the blood stream. Because these hormones have very short life spans there arises a constant need for these hormones when one either stands upright or experiences stress.

An additional process occurs during body movement that affects the blood pressure level. When an owner stands up or perceives stress, the sympathetic nerves discharge out of their endings. The endings of these nerves imbed in the blood vessels. These nerves release nor-epinephrine only, onto the motor end plate of the smooth muscles of the blood vessels. The release of nor-epinephrine causes these muscles to contract and thereby constrict the recipient blood vessels.

Nor-epinephrine released from the sympathetic nerves directly affects the blood vessel muscles contractile state. Concurrently nor-epinephrine released from the adrenal medulla into the blood stream diffuses towards the same muscular layer but from the other direction. The additive effect between the sympathetic nerve activation that releases nor-epinephrine into the blood stream and the adrenal secreted nor-epinephrine powerfully elevates blood pressure. The moderator of high blood pressure, in these situations, proves to be epinephrine. Epinephrine increases blood flow to the heart, muscles, and liver. Blood flow increases when blood vessels expand. Adequate epinephrine release therefore proves as one cornerstone for the prevention of high blood pressure.

Normally, this results in a tug-of-war between the full contraction and full relaxation of the blood vessel. Where a blood vessel ends up on the continuum is determined by the sum of the informational substances within the blood stream and the nervous tone instructions that reach this blood vessel. When a sub-optimal or unbalanced release of adrenal medulla derived epinephrine occurs then an increase in blood pressure becomes likely. Awareness for this process allows for a nutritional strategy that may be implemented to reduce blood pressure and the exacerbations of the symptoms of heart disease. After all, part of the heart disease experience concerns the diminished blood flow within the coronary arteries that a diminished epinephrine allows. Diseased coronary arteries are narrowed and hence need all the expansion they can receive from the help of epinephrine's message.

In order to obtain epinephrine the adequate molecular building parts and all of the nutritional cofactors must be present within the adrenal gland (discussed specifically below). These factors form a necessary pre-condition for the biosynthesis of epinephrine within the adrenal gland. Epinephrine and nor-epinephrine derive from adequate supplies of the amino acid tyrosine. Tyrosine can be obtained from the amino acid phenylalanine.

The adrenal gland cannot convert tyrosine to epinephrine unless all of the nutritional cofactors exist within the adrenal medulla (see below). Each step of the assembly line that eventually leads to the end product of epinephrine has a scientifically validated cofactor that must be present. If any one of these nutritional cofactors are missing, assembly stops and epinephrine synthesis becomes impossible.

The most common adrenal medulla deficiency involves the methyl donor system (explained below).

The methyl donor system proves necessary for the conversion of norepinephrine to epinephrine. The last step in the assembly process involves the conversion of norepinephrine to epinephrine. This last step cannot occur without a particular cofactor that disintegrates with each epinephrine molecule made. This is one of the cofactors vulnerable to deficiency. This cofactor is called S-Adenosyl methionine [SAMe for short]. SAMe is part of the methyl donor system. The SAMe cofactor donates its one methyl group to make one new epinephrine molecule from one norepinephrine. This spent cofactor needs to be recharged. If it is not recharged it is called S-adenosyl homocysteine. This deactivated molecule will further degrade into adenosine and homocysteine within the blood stream.

Unless adequate SAMe remains available in the body, adequate epinephrine biosynthesis cannot occur. Insufficient epinephrine biosynthesis leads to an increased biosynthesis of nor-epinephrine. Increased release of nor-epinephrine will raise the blood pressure. An additional exacerbating factor for blood pressure elevation exists when SAMe proves deficient.

When SAMe levels deplete a marked decrease in the body's ability to clear both epinephrine and nor-epinephrine from the blood stream arises. Not only does the body have the wrong hormone being secreted [nor-epinephrine] because of the SAMe deficiency, but because of this same deficiency there occurs the decreased ability to remove nor-epinephrine from the blood stream! These hormones clear from the body when other SAMe's in the circulation methylates them. The difference between norepinephrine and epinephrine is that the later contains a methyl group, which SAMe donates. However both hormones require additional methyl additions for their deactivation and removal from the body. When nor-epinephrine and epinephrine

receive these types of methyl additions they become inactive. Once inactivated they readily secrete from the kidney into the urine.

The SAMe deficient owner can become trapped in a vicious cycle of high blood pressure because their body's possess a diminished ability to manufacture epinephrine and from the decreased clearance of the sub-optimal nor-epinephrine. Without inactivation from other SAMe in the circulation this hormone remains free to continue spreading it's contraction message.

SAMe disintegrates within the body at the rate of one billion times a second (Cooney). The many other roles SAMe plays within healthy owners is explained in *The Body Heals*. For now realize that nutritional attention for the ways to recharge one's methyl donor system (SAMe is one of the members in this group) often has the ability to heal the epinephrine deficiency and the nor-epinephrine excess. The correction of epinephrine deficiency proves important because only epinephrine can oppose the clamping down of the sympathetic nerves during stressful times [see above discussion].

Deficient epinephrine output during adrenal medulla activation allows the blood pressure to rise dramatically.

The blood pressure rises dramatically because all information occurs as contraction information at the blood vessel level. This result occurs because only epinephrine moderates the contractile response. Epinephrine can do this because it contains message content, which directs the liver, heart and skeletal muscles blood vessels to dilate. Conversely, nor-epinephrine cannot do this because it contains all contractile message content. All blood vessels in the body, except the brain, are being directed to contract, therefore, the blood pressure rises dramatically. The same deficiency that diminishes synthesis of epinephrine also causes the diminished breakdown of elevated nor-epinephrine (see above). This added insult occurs because SAMe proves necessary to inactivate nor-epinephrine.

Only epinephrine contains the special message content for opening up the blood vessels within the heart, skeletal muscle, and liver. Sufficient epinephrine within the body directs blood into the above important areas and this fact prevents the dramatic rise in blood pressure. Blood pressure does not rise dramatically because the increased blood flow within these areas offsets the decreased blood flow elsewhere within the body. Within the rest of the body (excluding the brain) epinephrine directs the clamping down of blood vessels (constriction raises the pressure). The net effect usually evidences as a slight rise in the upper blood pressure value (systolic) and a slight lowering of the lower blood pressure reading (diastolic). Medical writings describe this as a widened pulse pressure. A widened pulse pressure indicates an increased cardiac output. In contrast, the stress response directed from the sympathetic nerves directs the constriction of all blood vessels except the brain. Adequate epinephrine release proves crucial in times of stress to prevent the sky rocketing of the blood pressure. Sky rocketing blood pressure means that blood flow delivery to heart, muscle and liver decreases. The sympathetic nerve activation coupled with the constriction effects of nor-epinephrine from the adrenal medulla will increase blood pressure abnormally, without adequate epinephrine.

The below listed cofactors are well documented in basic biochemistry textbooks as all being necessary for epinephrine to be manufactured. Basic medical physiology textbooks point out the marked difference between the effects on blood flow patterns and blood pressure between epinephrine and nor-epinephrine.

The synthetic sequence of catecholamines is: tyrosine-dopa- dopamine-nor epinephrine-epinephrine. The necessary cofactors that are needed in the synthesis sequence of tyrosine to the end product epinephrine are: tetrahydrobiopterin (made from folate), pyridoxal phosphate (vitamin B6), vitamin C, and SAMe.

Many additional cofactors and vitamins prove necessary for the recharging of SAMe. These cofactors and vitamins needed to recharge SAMe are called the methyl donor system. The molecules that make up the methyl donor system are: methionine, serine, vitamin B6, vitamin B12 and folate. These factors are involved in the remanufacture of SAMe once it has been degraded to S-adenosyl homocysteine. The methyl group, contained in SAMe, is needed to convert nor-epinephrine to epinephrine.

All of these cofactors involved in recreating SAMe, are known collectively as the **methyl** donor system. Depletion of this system has predictable consequences but paradoxically has been largely ignored by mainstream medicine in the clinical setting. It should be emphasized that the consumption of extra methionine or SAMe without proper attention to the adequacy of the other methyl donors will lead to elevated blood homocysteine levels because of the inability to recharge SAMe after creating each epinephrine. This will occur at the rate of one billion times a second when the deficiency proves severe.

Summary of vitamins and cofactors for the conversion of tyrosine to epinephrine

- Tetrahydrobiopterin [made from folate]
- Vitamin C
- Vitamin B6
- SAMe

A real food diet provides most of these cofactors and the methyl donor group, especially if one eats eggs. Eggs are rich in methionine. However, a processed food diet will likely lack one or more of these vitamins and cofactors. Many B-vitamin formulations are often deficient in folate content. Without adequate folate the methyl donor system will not function. All members of the methyl donor system need to be present or SAMe levels fall and homocysteine levels will rise.

The rise of blood homocysteine levels has been documented to signal a powerful risk factor for blood vessel disease. However, if one applies basic biochemical principles to the analysis for a homocysteine role in the development of heart disease it reveals it to be an unlikely agent in the direct injury of blood vessels. Rather it is more probable that it provides a biochemical red flag that nutritionally something is wrong. When blood homocysteine rises epinephrine synthesis will decrease proportionately. In *The Body Heals* it was explained how methyl also proves necessary for repair, brain fat manufacture, neurotransmitter formation, DNA stabilization and detoxification processes throughout the body. Remember that the body uses methyl at the rate of one billion times a second.

Elevated homocysteine levels denote a malfunction within the **methyl donor system** that leads to decreased epinephrine production and increased nor-epinephrine production. Elevated blood homocysteine levels may reflect a convenient biochemical marker to identify a depleted methyl donor system. One of the pathologies of a depleted methyl donor system is that it causes high blood pressure to develop (among other things). Elevated blood pressure and diminished blood flow to the heart muscle will result whenever nor-epinephrine occurs in the blood stream in higher than normal amounts compared to epinephrine. Dr. Steven Gordon of Whitefish, Montana, points out that finding an elevated blood homocysteine level may provide high blood pressure's etiology and it's solution as well.

It needs to be acknowledged that there are a group of owners who have high homocysteine levels without an elevation in their blood pressure. Some of these owners maintain a normal blood pressure because their bodies lower their anabolic and aldosterone steroids production. This compensation produces other aging effects (discussed in *The Body Heals*). Other owner's hearts produce high levels of the hormone, atrial natruretic peptide. This hormone overrides blood pressure elevation tendencies that other adrenal hormones promote. Both of these types of owners are found in the minority and hence blood pressure elevation occurring around middle age is the norm.

When one appreciates the fundamental role that both a highly functional and interrelated methyl donor system plays in body health, a way to treat blood pressure becomes possible. When this approach is considered in disease prevention strategies it begins to make sense why the adrenal medulla nutritional state is important.

This brings up the beauty of healing paths versus symptom control medicine. Prescriptions are all about symptom control and contain all the inevitable side effects as well. Healing does not have negative side effects. This is because once a problem is fixed it is over with.

None of these cofactors is more risky than if one takes of a multiple vitamin, in the general population. Each of the above listed cofactors has proven biochemical necessity in the synthesis of the epinephrine hormone. Healing involves working with the body to correct unbalanced states. Healthy owner's adrenals predictably contain optimal amounts of each of these cofactors and produce adequate epinephrine to maximize bodily function.

The caution is to avoid allergic reactions, which some owners have to the fillers and trace contaminants in certain brands of nutritional supplements. In general one should choose the best brand and quality (pharmaceutical grade). Once in a while there will be an owner who is allergic to vitamin C. Obtain the advice of a competent physician who will work on natural healing of the adrenal function and reduce the blood pressure naturally.

It turns out that the sympathetic tone is often increased in hypertensives and this central nervous system effect contributes greatly to the observed increase in blood pressure. The neurotransmitter involved in this case is nor-epinephrine. What is often under appreciated with regard to blood pressure is the tug of war between the sum of the hormones' message content within the blood stream and the message content delivered by the central nervous system. This dynamic equilibrium provides insight into the consequences of epinephrine deficiency.

Nor-epinephrine is a more powerful messenger [has more effect] when delivered within the nerves to the blood vessel (these nerves end in the muscular layer and when active direct contraction here) than when it is acting within the blood stream. Nor-epinephrine has more ability to raise blood pressure at a given concentration of secretion within the nerves than it does when it is secreted into the blood stream. This is a subtle but important point. Epinephrine deficiency can cause blood pressure elevation merely because there is an insufficient counter balance to the powerful nerve message contained in the presence of nor-epinephrine.

Epinephrine, within the blood vessels, has powerful effects starting at 50pg/ml, but nor-epinephrine doesn't exert its vasoconstrictor effects within the blood stream until 1500pg/ml. Nor-epinephrine's ability to raise blood pressure is mainly through its affect as a neurotransmitter within the sympathetic nervous system. Epinephrine is a counter response to the sympathetic nervous system's tendency to raise blood pressure.

Bibliography

Abou-Seif, MA., Youssef, AA. *Oxidative Stress and Male IGF-I, Gonadotropin and Related Hormones in Diabetic Patients.* Clin Chem Lab Med, July, Vol. 39, No. 7, 2001.

Abrams, William B., M.D., et al. *The Merck Manual of Geriatrics.* New Jersey: Merck Sharp & Dohme Research Laboratories, 1990.

Arvat, Emanuela, et. Al. (2000) Stimulatory Effect of Adrenocorticotropin on Cortisol, aldosterone and Dehydroepiandrosterone Secretion in Normal Humans: Dose Response Study. The Journal of Clinical Endocrinology and Metabolism. Vol. 85 No. 9 pages 3141-3146

Adkins, Robert C., MD. *Dr Adkins' New Diet Revolution.* 2nd ed. New York: M. Evans and Company, Inc, 1999.

_____. *Dr Adkins' Vita-Nutrient Solution.* New York: Simon and Shuster, 1998.

Aoki, Kazutaka, et al. Dehydroepiandrosterone Suppresses the Elevated Hepatic Glucose-6-Phosphatase and Fructose-1, 6-Biophosphatase Activities in C57BL/Ksj-db/db Mice. Diabetes, Vol. 48, August 1999.

Balch, James F. M.D., et al. *Prescription of Natural Healing.* New York: Garden City Park, 1990.

Barazzoni, R., et. Al. (2000) Increased Fibrinogen Production in Type 2 Diabetic Patients without Detectable Vascular Complications: Correlation with Plasma Glucagon Concentrations. Journal of Clinical Endocrinology and Metabolism. Vol. 85. No. 9 pages 3121-3125

Bareford, D., (1986) Effects of Hyperglycemia and sorbitol accumulation on erythrocyte deformability in diabetes mellitus. Journal of Clinical Pathology. Vol. 39 Issue 7

Becker, Robert O., MD, et al. *The Body Electric.* New York: William Morrow and Company, Inc, 1985.

Beers, Mark H., MD, ed. *The Merck Manual of Diagnosis and Therapy.* Whitehouse Station, NJ: Merck Research Laboratories, 1999.

Bellack, Leopold, MD, ed. *Psychology of Physical Illness.* New York: Grune & Stratton, 1952.

Bellamy, MF et Al. (1998) *Hyperhomocystinemia After an Oral Methionine Load Acutely Impairs Endothelial Function in Healthy Adults.* Circulation 98:1848-1852.

Bergman, Richard N., et al. Free Fatty Acids and Pathogenesis of Type 2 Diabetes Mellitus. Trends In Endocrinology and Metabolisim, 11, 2000.

Berkow, Robert, MD, ed. *The Merck Manual of Medical Information.* Whitehouse Station, NJ: Merck Research Laboratories, 1997.

Bernstein, Richard K., MD, F.A.C.E. *Diabetes Solution.* New York. Little, Brown and Company, 1997.

Bland, Jeffery S., PhD, ed. *Clinical Nutrition: A Functional Approach*. Gig Harbor, WA: Institute for Functional Medicine, 1999.

Bland, Jeffrey, Ph.D., *Nutritional Endocrinology*. Washington: Metagenics Educational Programs, 2002.

Bratman, Steven, M.D., et al. *Natural Health Bible*. 2nd ed. California: Prima Health, 2000.

Bricklin, Mark, ed. *The Practical Encyclopedia of Natural Healing*. Emmaus, PA: Rodale Press, 1976.

_____,et al. *The Practical Encyclopedia of Natural Healing* New, Revised Edition. Emmaus, PA: Rodale Press, 1983.

Brink, Marijke, et al. Angiotensin II Induces Skeletal Muscle Wasting Through Enhanced Protein Degradation and Down-Regulates Autocrine Insulin-Like Growth Factor I. Endocrinology, Vol. 142, No. 4, 2001.

Bruce, Debra F., et al. *The Unofficial Guide to Alternative Medicine*. New York: Macmillian, Inc, 1989.

Burr, Harold S. *Blueprint for Immortality*. *Essex*, England: The C.W. Daniel Company Limited, 1972.

Brand, Paul, MD, et al. *Fearfully and Wonderfully Made*. Grand Rapids, MI: Zondervan Publishing House, 1980.

Caine, Winston K., et al. *The Male Body: An Owner's Manual*. Emmaus, PA: Rodale Press, Inc, 1996.

Carey, Ruth, Ph.D., et al. *Common Sense Nutrition*. California: Pacific Press Publishing Association, 1971.

Cattaneo, L., et al. Characterization of the Hypothalamo-Pituitary-IGF-I Axis in Rats Made Obese by Overfeeding. Journal of Endocrinology, February, Vol. 148, No. 2, 1996.

Chopra, Deepak M.D. *Ageless Body, Timeless Mind*. New York: Harmony Books, 1993.

Chambers, John (1999) Demonstration of Rapid Onset Vascular Endothelial Dysfunction after Hyperhomocystinemia. Circulation 99:1156-1160.

Childe, Doc L., *The HeartMath Solution*. New York: HarperCollins Publishers, 1999.

Choi, Cheol S., et al. Independent Regulation of in Vivo Insulin Action on Glucose Versus K+ Uptake by Dietary Fat and K+ Content. Diabetes, Vol. 51, April, 2002.

Christ, Emanual R. et al. (1998) Dyslipidemia in adult Growth Hormone Deficiency and the Effect of GH Replacement Therapy. Trends in Endocrinology and Metabolism 9: 200-206

Clasey, JL., et al. *Abdominal Visceral Fat and Fasting Insulin are Important Predictors of 24-Hour GH Release Independent of Age, Gender, and Other Physiological Factors.* J Clin Endocrinol Metab, August, Vol. 86, No. 8, 2001.

Clemente, Carmine, Ph.D. *Anatomy A Regional Atlas of the Human Body.* Maryland: Urban & Schwarzenberg, 1981.

Cooney, Craig PhD. *Methyl Magic Maximum Health through Methylation* Andrew Mcmeal Publishing, Kansas City, Mo.

Cousins, Norman. *Anatomy of An Illness as Perceived By the Patient.* New York: Bantam Books, 1979.

Company, Merck &. *The Hypercholesterolemia Handbook.* Pennsylvania: Merck Sharp & Dohme, 1989.

Cush, Keneth and Ralph DeFronzo (2000) recombinant Human Insulin-Like Growth Factor 1 Treatment for 1 week Improves Metabolic Control in Type 2 Diabetes by Ameliorating Hepatic and Muscle Insulin Resistance. The Journal of Clinical Endocrinology and Metabolism. Vol. 85 No. 9 pages 3077-3084

Cusi, Kenneth, et al. Recombinant Human Insulin-Like Growth Factor I Treatment for 1 Week Improves Metabolic Control in Type 2 Diabetes by Ameliorating Hepatic and Muscle Insulin Resistance. The Journal of Clinical Endocrinology and Metabolism, Vol. 85. No. 9, 2000.

Danese, Mark D., (2000) Effect of Thyroxine Therapy on Serum Lipoproteins in Patients with Mild Thyroid Failure: a Quantitative Review of the Literature. Vol. 85 No. 9. Pages 2993-3001

Davenport, Horace W., DSc. *A Digest of Digestion.* 2nd ed. Chicago: Year Book Medical Publishers Inc, 1978.

DeBoer, H., et al. Changes in Subcutaneous and Visceral Fat mass During Growth Hormone Replacement Therapy in Adult Men. Int. Journal of Related Metabolic Disorders, June, Vol. 20, No. 6, 1996.

De Leo, Vicenzo. Effect of Metformin on Insulin-Like Growth Factor (IGF) I and IGF-Binding Protein I in Polycystic Ovary Syndrome. The Journal of Clinical Endocrinology & Metabolism, December, Vol. 85, No. 4, 2000.

Diamond, John W., M.D. *An Alternative Medicine Definitive Guide to Cancer.* California: Future Medicine Publishing, Inc., 1997.

Dowsett, M. (1999) Drug and hormone interactions of aromatase inhibitors. Endocrine Related Cancer 6 181-185

Ejima J et Al. (2000) relationship of HDL cholesterol and red blood cell filterability: cross-sectional study of healthy subjects. Clinical Hemorheological Microcirculation 22(1): 1-7

Erickson MD, Robert A (2001) Testosterone-Its Real Impact. Journal of Longevity Vol 7 No. 9

Fawcett, JP, (1994) Does cholesterol depletion have adverse effects on blood rheology? Angiology Volume 45 Issue 3

Ferril, William MD *The Body Heals*, Bridge Medical Publishers, Whitefish, Montana. (2003)

Ferril MD, William (1998) *Molecular Mechanisms of Biological Aging*. Medicine Tree

Ferril MD, William (1998) *The Adrenal Mystery*. Medicine Tree

Fottner, C., et el. Regulation of Steroidogenesis by Insulin-Like Growth Factors (IGFs) in Adult Human Adrenocortical Cells: IGF-I and, more Potently, IGF-II Preferentially Enhance Androgen Biosynthesis Through Interaction With the IGF-I Receptor and IGF-Binding Proteins. Journal of Endocrinol, September, Vol. 158, No. 3, 1998.

Frankel, Edward. DNA: *The Ladder of Life*. 2nd ed. New York: McGraw-Hill Book Company, 1979.

Frost, Robert A., Lang, Charles H. Differential Effects of Insulin-Like Growth Factor I (IGF-I) and IGF-Binding Protein-1 on Protein Metabolism in Human Skeletal Muscle Cells. Endocrinology, Vol. 140, No. 9, 1999.

Gaby, Alan R., MD, et al. *Nutritional Therapy in Medical Practice*. Kent, WA: Wright/Gaby Seminars, 1996.

Gangong, William F., MD. *Review of Medical Physiology*. 10th ed. Los Altos: Lange Medical Publications, 1981..

_____. *Review of Medical Physiology*. Los Altos: Lange Medical Publications, 1971

_____. *Review of Medical Physiology*. 19th ed. Stamford, CT: Appleton&Lange, 1999.

_____. *Review of Medical Physiology*. 20thed. McGraw-Hill Companies, Inc., 2001.

Gdansky, E., et al. Increased Number of IGF-I Receptors on Erythrocytes of Women with Polycystic Ovarian Syndrome. Clinical Endocrinal, August, Vol. 47, No. 2, 1997.

Gerber, Richard MD. *Vibrational Medicine*. Sante Fe: Bear and Company, 1986.

Gerras, Charles, ed. *The Complete Book of Vitamins*. Emmaus, PA: Rodale Press Inc, 1977.
_____, et al. *The Encyclopedia of Common Diseases*. Emmaus, PA: Rodale Press, Inc., 1976.

Giller, Robert M., MD, et al. *Natural Prescriptions*. New York: Ballentine Books, 1994.

Glowacki, Rosen CJ, et al. Sex steroids, The Inuslin-Like Growth Factor Regulatory System, and Aging Implications for the Management of Older Postmenopausal Women. J Nutr Health Aging, Vol.2, No. 1, 1998.

Gokce MD, Noyan, (1999) Long Term Ascorbic Acid administration reverses Endothelial Vasomotor Dysfunction in Patients with Coronary artery Disease. Circulation ;99 pages 3234-3240

Goldberg Group, Burton, ed. *Alternative Medicine The Definitive Guide*. Washington: Future Medicine Publishing, Inc., 1994.

Golden GA et Al. (1998) Steroid hormones partition to distinct sites in a model membrane bilayer: direct demonstration by small-angle X-ray diffraction.

Goodman, David, 'Soy toxins', press release

Goodman, Paul, *Compulsory Mis-education and the Community of Scholars*. New York: Vintage Books, 1962.

Golden GA et al (1999) Rapid and opposite effects of cortisol and estradiol on human erythrocyte Na+, K+-ATPase activity: relationship to steroid intercalation into the cell membrane. Life Science 65(12):1247-55

Gori, Francesca, et al. Effects of Androgens on the Insulin-Like Growth Factor System in an Androgen-Responsive Human Osteoblastic Cell Line. Endocrinology, Vol. 140, No. 12, 1999.

Graham, Ian M (June 11 1997) Plasma Homocysteine as a Risk Factor for Vascular Disease. JAMA Vol 27, No. 22

Grant Ph D, William (November 1998) the role of milk and sugar in heart disease. The American Journal of Natural Medicine.

Greenspan, Francis S.,MD, et al., eds. *Basic and Clinical Endocrinology*. Stamford, CT: Appleton&Lange, 1997.

Griffin, Tom, M.D., et al. *The Physicians Blueprint Feeling Good For Life*. Arizona: New Medical Dynamics Inc., 1983.

Grinspoon, Steven, et al. Effects of Androgen Administration on the Growth Hormone-Insulin-Like Growth Factor I Axis in Men with Aquired Immunodeficiency Syndrome Wasting. Journal of Clinical Endocrinology and Metabolism, Vol. 83, No. 12, 1998.

Gurnell, Eleanor M., (2001) Dehydroepiandrosterone replacement therapy. European Journal of Endocrinology. 145 pages 103-106

Guyton, Arthur C., M.D. *Textbook of Medical Physiology*. 7th ed. Pennsylvania: W.B. Saunders Company, 1986.

Halmos, Gabor, et. al. (2000) Human Ovarian Cancer Express Somatostatin Receptor. The Journal of Clinical Endocrinology and Metabolism. Vol. 85 No. 10 pages3509-3512

Hamel, Frederick G., et al. Regulation of Multicatalytic Enzyme Activity by Insulin and the Insulin-Degrading Enzyme. Endocrinology, Vol. 139, No. 10, 1998.

Handelsman, DJ, Crawford, BA. Androgens Regulate Circulating Levels of Insulin-Like Growth Factor (IGF)-I and IGF Binding Protien-3 During Puberty in Male Baboons. Journal of Clinical Metabolism, January, Vol. 81, No. 1, 1996.

Hanley, Anthony J.G., et al. Increased Proinsulin Levels and Decreased Acute Insulin Response Independently Predict the Incidence of Type 2 Diabetes in the Insulin Resistance Atherosclerosis Study. Diabetes, Vol. 51, April, 2002.

Hansten, Philip. *Drug Interactions*. 4th ed. London: Henry Kimpton Publishers, 1979.

Harper, Harold A., PhD. Review of Physiological Chemistry. 7th ed. Los Altos: Lange Medical Publications, 1959.

Harris, J.R., ed. *Blood Cell Biochemistry, Erythroid Cells*. New York: Plenum Press, 1990.

Harrington, James and Christin Carter-Su (2001) Signaling Pathways activated by the growth hormone receptor. Trends in Endocrinology. Vol. 12 No. 6 August 2001

Harrison, George R. *How Things Work*. New York: William Morrow and Co., 1941.

Hayes, Francis J., (2000) aromatase Inhibition in the Human Male Reveals a Hypothalamic Site of Estrogen Feedback. Journal of Clinical Endocrinology and Metabolism. Vol. 85 No. 9 pages 3027-3035

Heitzer, Thomas (2000) Tetrahydrobiopterin Improves Endothelium-Dependent Vasodialtion in Chronic Smokers. Circulation Research;86:e36

Heller, Richard F., MS, PhD, et al. *The Carbohydrate Addict's Healthy Heart Program*. New York: Ballentine Publishing Group, 1999.

Hendrickson, James E., MD. *The Molecules of Nature*. New York: W.A. Benjamin, 1965.

Hiramatsu R, and Nisula BC (1987 June) Erythrocyte-associated cortisol: measurement, kinetics of dissociation and potential physiological significance. Journal of Clinical Endocrinology and Metabolism. Vol. 64 No. 6 pages 1224-32

Hiramatsu, Ryoh and Bruce C. Nisula (1990) Uptake of erythrocyte-associated component of blood testosterone and corticosterone to rat brain. Journal of Steroid Biochemistry. Pages 383-87

Hiramatsu, R (1991) Uptake of erythrocytes-associated component of blood testosterone and corticosterone to rat brain. J of Steroid Biochemistry Mol Biol Mar 38: 383-7

Isaacson, Robert L., et al. *Toxin-Induced Blood Vessel Inclusions Caused by the Chronic Administration of Aluminum and Sodium Fluoride and Their Implications for Dementia*. Annals New York Academy of Sciences,

Jacobson GM (1975) 17 Beta-estradiol transport and metabolism in human red blood cells. J Clin Endocrinology and Metab. Feb 40 Issue 2

Jawetz, Ernest, MD, PhD, et al. *Review of Medical Microbiology*. 15th ed. Los Altos: Lange Medical Publications, 1982.

Glandular Failure-Caused Obesity

Jin, Weijun et Al. (2002) Lipases and HDL metabolism. Trends in Endocrinology Vol 13 No. 4 May 2002

Junqueira, Luis C., MD, et al. *Basic Histology.* 3rd ed. Los Altos: Lange Medical Publications, 1980.

Kamat, Amrita, et. Al. (2002) Mechanisms in tissue-specific regulation of estrogen biosynthesis in humans. Trends in Endocrinology and Metabolism. Vol. 13 April 2002 pgs. 122-128

Kellner, Michael, et al. Atrial Natriuretic Factor Inhibits the CRH-Stimulated Secretion of ACTH and Cortisol in Man. Life Sciences, Vol. 60, 1992.

Kemper, Donald ed. *Healthwise Handbook.* Idaho: Healthwise, Inc. 1976.

Keough, Carol, ed. *Future Youth.* Emmaus, PA: Rodale Press, Inc, 1987.

Khalsa, Dharma Singh, MD, et al. *Brain Longevity.* New York: Time Warner Company, 1997.

Kirpichnikov, Dmitri, and James Sowers (2001) Diabetes mellitus and diabetes-associated vascular disease. Trends in Endocrinology and Metabolism. Vol. 12 No. 5 July 2001.

Kishi, Yutaka, et al. Alph-Lipoic Acid: Effects on Glucose Uptake, Sorbitol Pathway, and Energy Metabolism in Experimental Diabetic Neuropathy. Diabetes, Vol. 48, October, 1999.

Klaassen, Curtis D., Ph.D. *Casarett & Doull's Toxicology.* 6th ed. McGraw-Hill Medical Publishing Division, 2001.

Klatz, Ronald, et al. *Grow Young with HGH.* New York: Harper Perennial, 1997.

Kotelchuck, David, ed. *Prognosis Negative.* New York: Vintage Books, 1976.

Kraemer W.J., et el. Effects of Heavy-Resistance Training On Hormonal Response Patterns In Younger VS. Older Men. Journal of Applied Physiology, September, Vol. 87, No.3, 1999.

Krupka RM and R Deves (1980) asymmetric binding of steroids to internal and external sites in the glucose carrier of erythrocytes. Biochim biophys Acta Vol 598 Issue 1

Lacayo, Richard (April 24, 2000) Testosterone. TIME Magazine: Page 58

Lasley, Bill L., et al. The Relationship of Circultating Dehydroepiandrosterone, Testosterone, and Estradiol to Stages of the Menopausal Transition and Ethnicity. The Journal of Clinical Endocrinology and Metabolism, Vol. 87, No. 8, 2002.

Laughlin, Gail and Elizibeth Barret-Conner (2000) Sexual dimorphism in the Influence of Advanced Aging on the Adrenal Hormone levels: The Rancho Bernardo Study. The Journal of Clinical Endocrinology and Metabolism pgs 3561-3568

Leavelle, Dennis E., MD, ed. *Mayo Medical Laboratories Interpretive Handbook.* Rochester, MN: Mayo Medical Laboratories, 1997.

Lee, John R., MD, et al. What *Your Doctor May Not Tell You About Premenopause*. New York: Warner Books, Inc, 1999.

_____. *Natural Progesterone: the Multiple Roles of a Remarkable Hormone*. Sebastopol, CA: BLL Publishing, 1993.

Lehninger, Albert L. *Biochemistry*. New York: Worth Publishers, Inc, 1975.

_____ *Biochemicstry*. New York: Worth Publishers, Inc., 2000.

Levitt, B.B. *Electromagnetic Fields*. New York: Harcourt Brace and Company, 1995.

Lewis, John G., et al. Caution on the use of saliva measurements to monitor absorption of progesterone from transdermal creams in postmenopausal women. Maturitas, 4, 2002.

Ley, Beth. DHEA: *Unlocking the Secrets to the Fountain of Youth*. California: BL Publications, 1996.

Lovern, J.A., The Chemistry of Lipids of Biochemistry Significance. London: Methuen & Co. LTD, 1955.

Lowe, John C., et al. *The Metabolic Treatment of Fibromyalgia*. Boulder, CO: Mc Dowell Publishing Company, 2000.

Lowenthal, Albert A., MD. *Endoctrine Glands and Sexual Problems*. Chicago:_____, 1928.

Lorand, Arnold, MD. *Old Age Deferred*. Philadelphia: F.A. Davis Publishers, 1911.

Martin, Janet L., et al. Insulin-Like Growth Factor Binding Protein-3 Is Regulated by Dihydrotestosterone and Stimulates Deoxyribonucleic Acid Synthesis and Cell Proliferation in LNCaP Prostate Carcinoma Cells. Endocrinology, Vol. 141, No. 7, 2000.

Mauras, Nelly, et. al. (2000) Estrogen Suppression in Males: Metabolic Effects. The Journal of Clinical Endocrinology and Metabolism. Vol85 No. 7 pages2370-2377.

McIntosh, M., et al. Opposing Actions of Dehydroepiandrosterone and Corticosterone in Rats. Proc Soc Exp Biol Med, July, Vol. 221, No. 3, 1999.

Mawatari S and Murakami K. (1999) Effects of ascorbic acid on peroxidation of human erythrocyte membranes by lipoxygenase. Ntrition Science vitaminology (Tokyo) Dec 45(6)687-99

McCann, Una D. (August 27, 1997)Brain Serotonin Neurotoxicity and Primary Pulmonry Hypertension From Fenfluramine and Dexfenfluramine. JAMA Vol.278, No 8

McCarty, MF. Androgenic Progestins Amplify the Breast Cancer Risk Associated with Hormone Replacement Therapy by Boosting IGF-I Activity. Med Hypotheses, February, Vol. 56, No. 2, 2001.

McCarty, MF. Modulation of Adipocyte Lipoprotein Lipase Expression as a Strategy for Preventing or Treating Visceral Obesity. Med Hypotheses, August, Vol. 57, No. 2, 2001.

McEvoy, Gerald K., Pharm.D, ed. *AHFS Drug Information*, 2001. Bethesda, MD: American Society of Health-System Pharmacists, Inc, 2001.

_____. *AHFS Drug Information*, 1986. Bethesda , MD: American Society of Health-System Pharmacists, Inc, 1986.

Mchedlishvili, G, New evidence for involvement of blood rheological disorders in rise of peripheral resistance in essential hypertension. Clinical Hemorheology Microcirculation Vol 17 Issue 1

McLaughlin, T, et Al. (2000) Carbohydrate Induced Hypertriglyceridemia: An Insight into the Link between Plasma Insulin and Triglyceride Concentrations. The Journal of Clinical Endocrinology and Metabolism. Vol85. No. 9 pages 3085-3088

Mellon, Cynthia H. and Lisa D. Griffin (2002) Neurosteroids: biochemistry and clinical significance. Trends in Endocrinology and Metabolism 13 pages 35-43

Mendelsohn, Robert S., M.D. *Confessions of a Medical Heretic*. New York: Warner Books Inc., 1979.

Mindell, Earl L., R.Ph.D, Ph.D., et al. *Dr. Earl Mindell's Secrets of Natural Health*. Illinois: Keats Publishing, 2000.

Mokken FC, et. Al. (1992) the clinical importance of erythrocyte deformability, a hemorheologically parameter. Annals of Hematology Volume 64 Issue 3

Morales, AJ. et al. The Effects of Six Months Treatment with a 100 mg Daily Dose of Dehyroepiamdrosterone (DHEA) on Circulating Sex Steroids, Body Composition and Muscle Strength in Age-Advanced Men and Women. Clinical Endocrinology (Oxf), October, Vol. 49, No. 4, 1998.

Morin, Laurie C., (2000) Endocrine and Metabolic Effects of Metaformin vs. Ethinyl-Cyproterone acetate in Obese Women with Polycystic Ovary Syndrome: A Randomized Study. The Journal of Clinical Endocrinology and Metabolism. Vol. 85 No. 9 pages 3161-3168

Morley J.E., et al. Potentially Predictive and Manipulable Blood Serum Correlatives of Aging in the Healthy Human Male: Progressive Decreases in Bioavailable Testosterone, Dehydroepiamdrosterone Sulfate, and the Ratio of Insulin-Like Growth Factor 1 to Growth Hormone. Pro Natl Acad Sci USA, July, Vol. 94, No.14, 1997.

Moss, Ralph W. *The Cancer Industry*. New York: Paragon House, 1989.

Monte, Tom, et al. *World Medicine*. New York: F.P. Putnam's Sons, 1993.

Mullenix, Phyllis J., Neurotoxicity of Sodium Fluoride in Rats. Neurotoxicology and Teratology, Vol. 17, No. 2, 1995.

Munzer, T., et al. Effects of GF and/or Sex Steroid Administration on Abdominal Subcutaneous and Visceral Fat in Healthy Aged Women and Men. J Clin Endocirinol Metab, August, Vol. 86, No. 8, 2001.

Muramoto, Naboru. *Healing Ourselves.* New York: Avon Books, 1973.

Murray, Michael, N.D., et al. *Encyclopedia of Natural Medicine.* California: Prima Health, 1998.

Myss, Caroline, PhD, et al. *Creation of Health.* New York: Three Rivers Press, 1993.

Nam, S.Y., et al. Low-Dose Growth Hormone Treatment Combined with Diet Restriction Decreases Insulin Resistance by Reducing Visceral Fat and Increasing Muscle Mass in Obese Type 2 Diabetic Patients. Int J Obes Relat Metab Disord, August, Vol. 25, No. 8, 2001.

Nelson, David L., et al. *Lehninger Principles of Biochemistry.* 3rd ed. New York: Worth Publishers, 2000.

Netzer, Corinne T. *Encyclopedia of Food Values.* New York: Dell   Publishing, 1992.

Nicklas, B.J., et al. Testosterone, Growth Hormone and IGF-I Response to Acute and Chronic Resistive Exercise in Men Aged 55-70 Years. Int. Journal of Sports Medicine, October, Vol. 16, No. 7, 1995.

Nitenberg A. Acetylcholine induced coronary vasoconstriction in young, heavy smokers with normal coronary arteriographic findings. Service d'Explorations Fonctionnelles, Unite 251, France.

Okada, Hidetaka, et. al. (2000) Progesterone Enhances Interleukin-15 Production in Human Endometrial Stromal Cells in Vitro. Journal of Clinical Endocrinology and Metabolism. Volume 85. No. 12 pages 4765-4770

Ornstein, Robert, et al. *The Amazing Brain.* Boston: Houghton Mifflin Company, 1984.

O'Rourke, P.J. *Parliament of Whores.* New York: The Atlantic Monthly Press, 1991.

Paolisso G., et al. Insulin Resistance and Advancing Age: What Role For Dehydroepiandrosterone Sulfate? Metabolism, November, Vol.46, No.11, 1997.

Pascal, Alana. *DHEA the Fountain of Youth Discovered?* California: Ben-Wal Printing, 1996.

Peeke, Pamela, MD, MPH. *Fight Fat After Forty.* New York: Penguin Group, 2000.

Persson SU (1996) Correlations between fatty acid composition of the erythrocyte membrane and blood rheology data. Scandinavian Journal of Clinical Laboratory Investigation. April 96 vol. 56 Issue 2

Pert, Candace B., PhD. *The Molecules of Emotion.* New York: Simon and Schuster, Inc, 1997

Petersdorf, Robert G., M.D., et al. Harrison's *Principles of Internal Medicine* tenth edition. McGraw-Hill Book Company, 1983.

Pinchera, Aldo, MD, ed. *Endocrinology and Metabolism.* London: McGraw-Hill International(UK) Ltd., 2001.

Pino, Ana M. et. al. (2000) Dietary Isoflavones Affect Sex Hormone Globulin levels in Postmenopausal Women. The Journal of Clinical Endocrinology and Metabolism. Vol 85. No.8 pages 2797-2800

Pries, Axel R., et al. *Structural Autoregulation of Terminal Vascular Beds. Hypertension,* 1999.

Rath, Matthias, MD. *Eradicating Heart Disease.* San Francisco: Health Now, 1993.

Ravaglia, G., et al. Regular Moderate Intensity Physical Activity and Blood Concentrations of Endogenous Anabolic Hormones and Thyroid Hormones in Aging Men. Mech Aging Dev, February, Vol. 122, No. 2, 2001.

Ravel, Richard, MD. *Clinical Laboratory Medicine.* 6th ed. St Louis: Mosby-Year Book, Inc, 1995.

Raynaud-Simon, A., et al. Plasma Insulin-Like Growth Factor I Levels in the Elderly: Relation to Plasma Dehydroepiandrosterone Sulfate Levels, Nutritional Status, Health and Mortality. J Gerontology, July-August, Vol.47, No. 4, 2001.

Reaven, Gerald, M.D., et al. *Syndrome X.* New York: Simon & Schuster, 2000.

Reiss, Uzzi, M.D., et al. Natural *Hormone Balance for Woman.* New York: Pocket Books, 2001.

Remington, Dennis, M.D., et al. *Back to Health.* Utah: Publishers Press, 1986.

Rifkind, Richard, et al. *Fundamentals of Hematology.* 2nd ed. Illinois: Year Book Medical Publishers, Inc. 1980.

Robbins, John. *Reclaiming Our Health: Exploding the Myth and Embracing the Source of True Healing.* Tiburon, CA: HJ Kramer Inc, 1998.

Robbins, Stanley L., MD, et al. *Pathologic Basis of Disease.* 2nd ed. Philadelphia: W.B. Saunders Company, 1979.

Rodale, J.I., et al. *The Health Seeker.* Emmaus,PA: Rodale Books, Inc. 1972.

Rodale, J.I., ed. *Health Builder.*Pennsylvania: Rodale Press, Inc., 1971.

Roggenkamp HG (1986) Erythrocyte rigidity in healthy patients and patients with cardiovascualr disease risk factors. KWH Oct 1986 64: 1091-6

Rojo ND, Ruth (2001) Why is it harder to lose weight as we age? Journal of Longevity Vol 7 No 9

Rosedale MD, Ron Presentation at the Health Institute's boulder-Fest, August 1999 seminar

Rosenfeld, Isadore, M.D. *The Complete Medical Exam.* New York: Simon & Schuster, 1978.

Rosmond, Roland, et al. Stress-Related Cortisol Secretion in Men: Relationships with Abdominal Obesity and Endocrine, Metabolic and Hemodynamic Abnormalities. Journal of Clinical Endocrinology and Metabolism, February, Vol. 83, No. 6, 1998.

Rosmond, R, Bjortorp, P. The Interactions Between Hypothalamic-Pituitary-Adrenal Axis Activity, Testosterone, Insulin-Like Growth Factor I and Abdominal Obesity with Metabolism and Blood Pressure in Men. Int Journal Obes Relat Metab Disord, December, Vol. 22, No. 12, 1998.

Rubin, Philip, ed. *Clinical Oncology* sixth edition. American Cancer Society, 1983.

Ruiz, Gomez F. (1998) Treatments with progesterone analogues decreases macrophage Fcgamma receptors expression. Clinical Immunopathology Dec; 89(3): 231-9

Russell, A.L. Glycoaminoglycan (GAG) Deficiency in Protective Barrier as an Underlying, Primary Cause of Ulcerative Colitis, Crohn's Disease, Interstitial Cystitis and Possibly Reiter's Syndrome. Medical Hypotheses, Vol. 52, No. 4, 1999.

Ryan, Graeme B., MB, BS, PhD, et al. *Inflamation*. Kalamazoo, MI: The Upjohn Company, 1977.

Sapolsky, Robert M. *Stress, the Aging Brain, and the Mechanisms of Neuron Death*. Cambridge, MA: The MIT Press, 1992.

_____. *The Trouble with Testosterone*. New York: Simon and Shuster, Inc, 1997.

Sarno, John E., MD. *The Mindbody Prescription*. New York: Warner Books, Inc, 1998.

Simpson, Leslie O. (1987) Red cell and hemorheological changes in multiple sclerosis. Pathology, 19 pp51-55

Secomb, T. W. (1998) A model for red cell motion in glycocalyx-lined capillaries. American Journal of Physiology 274 H1016-H1022

Sahelian MD, Ray (October 1996) DHEA *Youth in a Bottle? Lets Live*

Schechter MD, Michael, et. Al. (2000) Oral Magnesium Therapy Improves Endothelial Function in Patients with Coronary Artery Disease. Circulation Nov. 7 2000. Pages 2353-2358

Scholl, B.F., PhD, MD, ed. *Library of Health*. Philadelphia: Historical Publishing, Inc, 1932.

Schwarzbein, Diana, MD, et al. *The Schwarzbein Principle*. Deerfield Beach, FL: Health Communications, Inc, 1999.

Sears, Barry, PhD, et al. *Enter the Zone*. New York: HarperCollins Publishers, Inc, 1995.

Sheally, C.N., MD, PhD, ed. *The Complete Family Guide to Alternative Medicine*. New York: Barnes&Noble, Inc, 1996.

Shippen, Eugene, MD, et al. *The Testosterone Syndrome*. New York: M.Evans and Company, Inc, 1998.

Signorello, LB., et al. Hormones and Hair Patterning In Men: A Role for Insulin-Like Growth Factor 1? Journal of the American Academy of Dermatology, February, Vol. 40, No. 2, 1999.

Sobel, David S., MD, et al. *The People's Book of Medical Tests*. New York: Simon and Schuster, 1985.

Solerte, Sebastiano Bruno, et al. Dehydroepiandrosterone Sulfate Enhances Natural Killer Cell Cytotoxicity in Humans Via Locally Generated Immunoreactive Insulin-Like Growth Factor I. The Journal of Clinical Endocrinology & Metabolism, Vol. 84, No. 9, 1999.

Spector, Walter G. *An Introduction to General Pathology*. 2nd ed. Edinburgh, Scotland: Churchill Livingstone, 1980.

Stelfox, Henry Thomas, M.D., et al. Conflict of Interest in the Debate Over Calcium-Channel Antagonists. The New England Journal of Medicine, January 8, 1998.

Stewart, Paul M., Tomlinson, Jeremy W. Cortisol, 11beta-Hydroxysteroid Dehydrogenase Type 1 and Central Obesity. TRENDS in Endocrinology & Metabolism, Vol. 13, No. 3, April, 2002. pgs 94-96

Stuart, J. (1985) Erythrocyte Rheology. J Clinical Pathology Vol. 38 Issue 9

Study Links high Carb to Cancer, Associated Press (April 2002)

Takaya, Kazuhiko, et. al. (2000) Ghrelin Strongly Stimulates Growth Hormone (GH) Release in Humans. The Journal of Clinical Endocrinology and Metabolism. Vol. 85. No. 12 pages 4908-4911

Thomas, Lewis, *The Lives of a Cell*. New York: Bantam Books, 1974.

Thrailkill, K.M. Insulin-Like Growth Factor-I in Diabetes Mellitus: its Physiologic, Metabolic Effects, and Potential Clinical Utility. Diabetes Technol Ther, Spring, Vol. 2, No. 1, 2000.

Tiller, William A. (1996) Cardiac Coherence: A New, Noninvasive Measure of Autonomic Nervous System Order. Alternative Therapies Jan 96, Vol. 2, No 1

Tissandier, O., et al. Testosterone, Dehydroepiandrosterone, Insulin-Like Growth Factor 1, and Insulin in Sedentary and Physically Trained Aged Men. Eur J Appl Physiol, July, Vol.85, No. 1-2, 2001.

Tsuda K et al (2001) Electron paramagnetic resonance investigation on modulatory effect of 17 Beta-estradiol on membrane fluidity of erythrocytes in postmenopausal women. Arteriosclerosis Thromb Vasc Biol Aug:21(8):1306-12

Tsuji, K. Specific Binding and Effects of Dehroepiandrosterone Sulfate (DHEA-S) on Skeletal Muscle Cells: Possible Implication for DHEA-S Replacement Therapy in Patients With Myotonic Dystrophy. Life Science, Vol. 65, No.1, 1999.

VanHaaften, M., et al. Identification of 16-alpha Hydroxyestrone as a Metabolite of Estriol. Gynecol, Endorinol 2, 1988.

Veldhuis, Johannes D., et al. Estrogen and Testosterone, But Not a Nonaromatizable Androgen, Direct Network Integration of the Hypothalamo-Somatotrope (Growth Hormone)-Insulin-Like Growth Factor I Axis in the Human: Evidence from Pubertal Pathophysioogy and Sex-Steroid Hormone Replacement. Journal of Clinical Endocrinology and Metabolism, Vol. 82, No.10, 1997.

Vendola, K., et al. Androgens Promote Insulin-Like Growth Factor-I and Insulin-Like Growth Factor-I Receptor Gene Expression in the Primate Ovary. Hum Reprod, September, Vol. 1, No. 9, 1999.

Viveiros, M.M., Liptrap, R.M. ACTH Treatment Disrupts Ovarian IGF-I and Steroid Hormone Production. Journal of Endocrinology, 164, 2000.

Volek J.S., et al. Body Composition and Hormonal Responses to a Carbohydrate Restricted Diet. Metabolism, July, Vol. 51, No. 7, 2002.

Vondra, K., et al. Role of the Steroids, SHBG, IGF-I, IGF BP-3 and Growth Hormone in Glucose Metabolism Disorders During Long-Term Treatment with Low Doses of Glucocorticoids. Cas Lek Cesk, February, Vol. 141, No. 3, 2002.

Wallach, Jacques, MD. *Interpretation of Diagnostic Tests*. 6th ed. New York: Little, Brown and Company, 1996.

Weast, Robert C., PhD. *Handbook of Chemistry and Physics*. 56th ed. Cleveland, OH: CRC Press, Inc, 1975.

Weil, Andrew, MD. *Natural Health, Natural Medicine*. Boston: Houghton Mifflin Company, 1990.
_____. *Health and Healing*. New York: Houghton Mifflin Company, 1995.

_____. *Spontaneous Healing*. New York: Alfred A. Knopf, Inc, 1995.

_____. *Eating Well for Optimum Health*. New York: Alfred A Knopf, 2000.

Whitaker, Julian MD. *Dr Whitaker's Guide to Natural Healing*. Rocklin, CA: Prima Publishing, 1995.

Whitaker MD, Julian (September 1998) DHEA helps regulate the immune system. Health and Healing Vol 8 No 9

Wild, Russell, ed. *The Complete Book of Natural and Medicinal Cures*. Emmaus, PA: Rodale Press, 1994.

Wilson, Helen E and Ann White (1998) Prohormone: their Clinical Relevance. Trends in Endocrinology and Metabolism 9:396-402

Wood, Ian (2002) Pro-inflammatory mechanisms of a nonsteroidal anti-inflammatory drug. Trends in Endocrinology Vol 13 No. 2 March 2002

Wright, Jonathan V., MD, et al. Natural *Hormone Replacement for Women Over 45*. Petaluma, CA: Smart Publications, 1997.

_____, et al. Natural *Hormone Replacement*. California: Smart Publications, 1997.

_____, et al. *The patient's Book of Natural Healing*. California: Prima Health, 1999.

Wright MD, Jonathan and Alan Gaby MD (October 16-19 1998) Nutritional Therapy in Medical Practice, Doubletree Seattle Airport Hotel

Yen, SS, Laughlin GA. Aging and the Adrenal Cortex. Exp. Gerontol, Nov-Dec, Vol. 33, No. 7-8, 1998.

Youl, Kang H., et al. Effects of Ginseng Ingestion on Growth Hormone, Testosterone, Cortisol, and Insulin-Like Growth Factor I Responses to Acute Resistance Exercise. J Strength Cond Res, May, Vol. 16, No. 2, 2002.

Zachrisson, I., et al. Determinants of Growth in Diabetic Pubertal Subjects. Diabetes Care, August, Vol. 20, No. 8, 1997.

Zager PG et Al. (1986) Distribution of 18-hydroxycorticosterone between red blood cells and plasma. J Clin Endocrinology Metab Jan 62: 84-9

Zborowski, Jeanne V., et. Al. (2000) Bone Mineral Density, Androgens, and the Polycystic Ovary: The Complex and Controversial Issue of Androgenic Influence in the Female Bone. Vol. 85 No. 10 pages 3496-3506

Index

24-hour urine 37, 39, 83, 180, 223, 234

absorption 61, 85, 135, 136, 151, 152, 157, 159, 162, 184, 191, 192, 194, 196, 199, 200, 201, 206, 207, 209, 290

acetate 22, 188, 197, 243, 245, 246, 248, 249, 291

acidophilus 208

ACTH 127, 289, 296

acute phase reactants 103, 108, 192, 194

adrenal cortex 119, 126, 272

adrenal insufficiency 119

adrenal medulla 66, 67, 68, 272, 273, 274, 275, 277, 278, 281

adrenaline 11, 75, 97, 119, 120, 121, 271

aging, 16, 19, 28, 34, 36, 58, 75, 76, 132, 144, 147, 161, 162, 184, 186, 212, 264, 280

albumin 187

aldosterone 18, 49, 50, 92, 93, 94, 124, 125, 126, 127, 132, 133, 147, 148, 258, 280, 283

alkaline 198, 204, 206

amino acids 16, 41, 63, 70, 120, 121, 154, 155, 159,

161, 162, 163, 205, 215, 219, 220, 221, 249, 264, 265, 266

anabolic 34, 39, 59, 60, 61, 105, 136, 137, 139, 140, 142, 144, 145, 280

androgen 26, 33, 34, 35, 36, 37, 40, 41, 43, 59, 63, 78, 80, 83, 85, 97, 99, 100, 103, 116, 117, 131, 139, 142, 161, 175, 179, 180, 210, 220, 226, 238, 254, 255

androstenedione 142

angiotensin converting enzyme 127, 148

angiotensin type two 126

antioxidants 136, 214, 215

anxiety 14, 88

appetite 12, 61, 75, 80, 81, 91, 98, 153, 167, 175

arachidonic acid 164, 267, 268

arginine 63, 215, 269

arthritis 21, 159, 164, 262, 268

asthma 262, 266

ATP 23, 185, 186, 188, 245, 246, 250

autoimmune disease 127, 159

Bernstein MD, Richard K, 81, 166, 283

bile 203, 215, 216, 239

biogenic amines 266

birth control pills 78, 115, 116

blood vessels 65, 66, 68, 89, 101, 103, 105, 108, 127, 134, 160, 168, 169, 172, 173, 176, 177, 189, 192, 194, 196, 210, 216, 226, 234, 235, 238, 239, 243, 266, 272, 273, 274, 277, 279, 282

bones 13, 19, 27, 40, 45, 91, 141, 189

bradykinin 127

brain 12, 33, 48, 59, 61, 81, 98, 127, 141, 144, 165, 167, 175, 189, 230, 245, 272, 273, 277, 279, 288

breast cancer 118, 252, 254, 258

calcium 45, 51, 52, 53, 54, 55, 56, 57, 86, 95, 124, 147, 155, 184, 185, 186, 187, 188, 189, 190, 191, 242

capillaries 127, 294

carbohydrates 12, 47, 48, 70, 73, 91, 95, 96, 97, 98, 99, 157, 158, 167, 168, 172, 175, 196, 197, 204, 208, 230, 256, 257

carbon dioxide 67, 204, 244, 246, 248

carnitine 244, 245, 246

cartilage 70, 160

catabolic 27, 34, 35, 39, 40, 41, 42, 59, 60, 105, 139, 140, 141, 142, 220, 234, 265

cell membrane 23, 24, 93, 144, 147, 155, 164, 165, 166, 176, 186, 187, 188, 190, 236, 250, 287

cell power plants 17, 24, 84, 185, 218, 235, 245, 246

cells 16, 17, 18, 19, 21, 22, 23, 24, 25, 26, 27, 28, 29, 30, 31, 32, 34, 36, 38, 39, 40, 42, 43, 44, 46, 47, 48, 49, 51, 59, 61, 70, 73, 74, 75, 76, 84, 87, 88, 89, 90, 91, 92, 93, 97, 100, 103, 105, 106, 108, 111, 112, 114, 115, 119, 122, 123, 127, 131, 139, 142, 143, 144, 145, 146, 147, 148, 151, 153, 155, 156, 157, 160, 161, 162, 163, 164, 165, 166, 168, 169, 170, 171, 172, 176, 177, 180, 183, 185, 186, 188, 189, 190, 191, 192, 198, 203, 204, 205, 207, 208, 209, 210, 211, 214, 215, 216, 217, 218, 219, 220, 221, 228, 230, 231, 232, 233, 235, 236, 237, 240, 244, 245, 247, 248, 249, 250, 251, 253, 263, 265, 266, 267, 269, 288, 297

chelation 192

chief cells 198

choline 164

chronic fatigue 37, 249

coenzyme A 246

coenzyme Q10 247, 248

colon 207, 208, 209, 210, 211, 213, 215, 216, 236, 238

constipation 184, 209, 210

coronary arteries 172, 274

cortisol 11, 30, 37, 40, 41, 42, 43, 60, 61, 63, 70, 75, 76, 79, 81, 83, 84, 94, 96, 97, 99, 102, 105, 106, 107, 108, 112, 116, 119, 127, 132, 138, 139, 140, 141, 142, 143, 175, 179, 187, 209, 210, 217, 219, 222, 223, 224, 225, 234, 237, 242, 244, 249, 272, 287, 288

C-peptide 102, 111, 113, 232, 255, 257

C-reactive protein 108, 192

depression 268

DHEA 26, 33, 34, 36, 42, 43, 59, 63, 97, 99, 111, 114, 122, 127, 138, 139, 142, 144, 210, 220, 221, 234, 240, 290, 291, 292, 294, 295, 296

DHT 60

diabetes 10, 14, 21, 32, 63, 75, 76, 80, 85, 88, 89, 104, 159, 169, 211, 212, 216, 219, 221, 225, 226, 228, 229, 231, 232, 233, 234,

237, 238, 254, 255, 262, 283, 289

DNA program 18, 19, 23, 24, 26, 34, 35, 36, 38, 39, 59, 122, 124, 126, 148, 211, 220, 221, 222, 236, 240, 250, 251, 263, 264

dopamine 65, 68, 75, 121, 240, 278

duodenum 201, 204, 205, 206, 207

emotions 5, 108, 137, 138, 139, 142, 143, 144, 146, 192

endoplasmic reticulum 250

endothelin 106, 266

enzyme machines 152, 153, 196, 198, 203, 265

epinephrine 11, 30, 40, 42, 61, 65, 66, 67, 68, 69, 70, 75, 76, 81, 96, 107, 112, 119, 120, 140, 217, 219, 222, 224, 225, 244, 249, 266, 271, 272, 273, 274, 275, 276, 277, 278, 279, 280, 281, 282

esophagus 197, 199, 200, 201

essential fatty acids 164, 165, 196, 267, 268

estrogen 11, 35, 36, 63, 78, 81, 84, 96, 100, 114, 115, 116, 117, 118, 119, 137, 180, 226, 237, 240, 242, 244, 289

exercise 10, 28, 37, 45, 60, 63, 67, 68, 74, 75, 76, 80,

83, 96, 97, 98, 108, 142, 144, 145, 146, 172, 174, 175, 179, 180, 218, 219, 228, 230, 231, 233, 240, 244, 253, 255, 257

exiting the torture chamber 95, 133

fats 12, 70, 96, 124, 144, 153, 163, 164, 165, 168, 176, 177, 178, 195, 196, 201, 204, 206, 267, 268, 269

ferritin 108, 192

fiber 184, 208, 209

fibrocystic breast disease, 118

fibroids 118

fluoride 105, 176, 214

folate 67, 69, 121, 134, 245, 270, 278, 279

fuel 16, 17, 21, 22, 24, 25, 26, 27, 28, 30, 31, 32, 33, 34, 41, 42, 61, 63, 65, 66, 67, 68, 70, 72, 74, 76, 83, 84, 98, 99, 103, 104, 108, 111, 112, 113, 115, 119, 122, 123, 135, 139, 140, 141, 143, 150, 151, 152, 156, 157, 165, 167, 170, 171, 172, 179, 188, 190, 194, 211, 212, 213, 216, 217, 218, 219, 220, 221, 228, 233, 235, 236, 244, 246, 248, 249, 250, 265, 266

galactosamine 160, 201

galactose 157, 196

gallbladder 203, 205, 206, 207

genetic programs 18, 59

GERD 200

ginseng 131, 144, 297

glandular secretions 21, 253, 255

glucosamine 160, 201

glucuronic 215

glutathione 215

glycogen 17, 25, 48, 70, 72, 89, 90, 167, 171, 217, 218, 219, 230, 231, 235, 265

glycosaminoglycans (GAG) 160

gonads 34, 35, 40, 60, 93, 96, 98, 99, 108, 117, 127, 131, 132, 136, 145, 148, 175, 179, 258

Gordon ND, Steven 280

ground substance 160

growth hormone 11, 18, 21, 26, 27, 28, 29, 30, 31, 33, 41, 42, 43, 60, 61, 63, 64, 70, 73, 74, 75, 76, 78, 80, 83, 84, 96, 97, 100, 103, 106, 107, 108, 111, 112, 115, 123, 140, 141, 142, 143, 161, 210, 217, 219, 220, 221, 222, 224, 228, 232, 233, 235, 236, 237, 240, 249, 253, 254, 255, 259, 288

heart disease 10, 14, 21, 80, 107, 164, 165, 169, 171, 172, 173, 175, 176, 192, 211, 212, 216, 219, 224, 225, 226, 229, 232, 234, 237, 238, 248, 256, 262, 268, 271, 274, 279, 287

heartburn 199, 200, 201, 202

hemoglobin 191, 192, 207, 239

hemosiderosis 191

hierarchy of hormones 60, 65, 165, 222, 263

histamine 65, 127, 199, 203

HMG Co A reductase 169, 228, 229

homocysteine 18, 69, 121, 258, 275, 278, 279, 280, 287

hormonal fats 164, 165, 201, 267, 268, 269

hormone imbalances 82, 85, 263

hormone mimics 84, 118, 119, 136, 137, 252

hormone mimics 117

hormones 11, 12, 13, 16, 17, 18, 19, 21, 22, 24, 25, 26, 27, 28, 30, 31, 33, 34, 35, 36, 37, 38, 39, 41, 42, 45, 49, 59, 60, 61, 63, 65, 66, 67, 70, 76, 79, 80, 81, 82, 85, 86, 87, 96, 97, 99, 100, 102, 108, 111, 112, 115, 117, 119, 123, 124, 131, 138, 139, 140, 141, 142, 143, 144, 145, 147, 150, 151, 152, 158, 164, 165, 167, 174, 179, 180, 181, 182, 185, 186, 187, 199, 202, 203, 209, 213, 216, 217, 218, 219, 220, 222, 223, 224, 225, 233, 240, 242, 249, 251, 252, 253, 254, 257, 258, 259, 260, 262, 263, 264, 265, 266, 267, 268, 269, 270, 272, 273, 276, 280, 281, 287

hydrochloric acid 202

hypoglycemia 175, 223

hypothalamus 75, 79

ileum 183, 205, 207

immune system 108, 159, 161, 165, 172, 192, 196, 216, 247, 296

informational substance 138, 147, 152, 155, 156, 162, 163, 164, 165, 180, 199, 262, 274

insulin like growth factor type 1 18, 25, 26

insulin receptor 30, 61, 70, 112, 221, 240, 253

insulin resistance 27, 28, 74, 75, 76, 84, 88, 89, 90, 103, 104, 157, 158, 225, 227, 230, 237, 240

jejunum 183, 205

joints 40, 77, 141

kidneys 45, 215, 216, 249

lactic acid 67, 135, 228, 248, 249

LDL cholesterol 33, 73, 75, 76, 90, 97, 98, 101, 168, 169, 170, 172, 173, 175, 177, 178, 216, 217, 219, 221, 225, 227, 238, 240, 244, 265

leukotrienes 164, 267

licorice root 202

ligaments 155

linoleic acid 164, 268

lipids 163, 164, 166, 290

lipoic acid 135, 214, 248

lipoprotein lipase 170

low voltage cell syndrome 93, 158

lungs 19, 127

lymphatics 171

macrophages 33, 73, 168, 169, 171, 172, 173, 177, 217, 219, 225, 227

mast cells 127, 170

message content 16, 18, 19, 21, 23, 24, 25, 28, 30, 31, 32, 33, 34, 35, 36, 37, 38, 39, 40, 41, 59, 60, 61, 63, 68, 70, 73, 80, 83, 84, 90, 91, 92, 94, 97, 99, 103, 104, 105, 108, 112, 115, 117, 119, 123, 125, 126, 127, 132, 137, 138, 139, 140, 141, 142, 144, 145, 148, 150, 157, 162, 164, 165,

167, 170, 175, 179, 187, 191, 199, 213, 217, 218, 219, 220, 221, 222, 224, 225, 227, 229, 231, 233, 234, 236, 237, 238, 239, 240, 251, 252, 253, 263, 264, 265, 266, 267, 269, 271, 272, 273, 277, 281

methionine 67, 121, 245, 275, 278, 279, 283

methyl donor system 68, 121, 245, 257, 271, 275, 276, 278, 279, 280

mitochondria 131, 132, 153, 185, 186, 188, 217, 236, 244, 245, 246, 247, 248, 250

mucopolysaccharides 160

mucous 200, 201, 202, 203, 205

muscles 13, 27, 29, 41, 42, 43, 59, 60, 61, 64, 65, 66, 67, 68, 70, 73, 74, 77, 91, 92, 113, 135, 141, 175, 218, 220, 222, 228, 230, 231, 232, 240, 248, 272, 274, 277

neurotransmitters 152

niacin 243, 244, 245, 246, 248

nitric oxide 106, 127, 257, 267, 269

nonsuppressible insulin like activity of the blood 161, 210, 253

noradrenaline 271

norepinephrine 121, 224, 271, 275, 276

osteoporosis 37, 40, 189

oxidants 136, 214, 215

pantothenic acid 134, 135, 214, 244, 245, 246, 248, 249

parietal cells 198

pepsin, 198, 202

pepsinogen 198, 202

phenylalanine 66, 69, 120, 275

phosphate 164, 189, 278

phospholipids 214

pituitary gland 26, 115

placenta 77, 78, 84, 116

portal vein 30, 31, 103, 112, 141, 157, 172, 216, 221, 234

potassium 12, 44, 45, 46, 47, 48, 49, 50, 51, 73, 86, 87, 88, 89, 90, 91, 93, 94, 95, 97, 100, 104, 124, 125, 127, 133, 158, 174, 175, 184, 185, 188, 218, 219, 226, 230, 231, 240

pregnenolone 124, 131, 132

progesterone 78, 116, 118, 123, 142, 144, 242, 258, 290, 294

prolactin 11, 21, 77, 78, 79, 116, 258

prostaglandins 164, 201, 267

protein 12, 22, 28, 29, 30, 31, 34, 42, 46, 60, 61, 63, 65, 70, 75, 76, 82, 83, 85, 87, 91, 92, 93, 96, 98, 107, 108, 112, 113, 117, 120, 139, 140, 142, 143, 153, 154, 155, 156, 157, 158, 159, 161, 162, 163, 165, 170, 172, 174, 175, 179, 180, 188, 189, 190, 192, 194, 198, 199, 200, 202, 204, 205, 206, 208, 215, 220, 221, 222, 224, 228, 229, 231, 234, 239, 242, 244, 246, 257, 268, 284, 285, 286, 290

PSA 255, 258

reverse T3 258

SAMe 67, 69, 121, 245, 275, 276, 277, 278, 279

Schwarzbein MD, Diana 80, 81, 166, 294

selenium 192, 214

serotonin 65, 75, 77, 78, 240, 290

sex hormone binding globulin, 242

sodium 12, 44, 45, 46, 48, 49, 50, 51, 52, 53, 54, 55, 56, 57, 82, 86, 87, 88, 93, 94, 95, 97, 104, 124, 127, 133, 158, 174, 184, 185, 188, 230, 231, 288, 291

steroid pressure 40, 127, 141

steroid tone 39, 126, 127, 131, 132, 133, 134, 136,

137, 138, 139, 140, 141,
142, 144, 145, 146, 159

steroids 18, 33, 34, 35, 36,
37, 38, 39, 40, 43, 49, 50,
59, 60, 61, 77, 78, 92, 119,
125, 126, 127, 131, 132,
134, 136, 137, 138, 140,
141, 142, 143, 144, 145,
155, 169, 180, 215, 224,
242, 258, 263, 272, 280,
286, 289, 291, 296

stomach 120, 136, 192, 197,
198, 199, 200, 201, 202,
203, 204, 205, 206, 238

stress hormone 37, 40, 45,
83, 104, 139, 144

sulfation factor 30, 160,
161, 201, 210, 238

survival response 108, 139

Syndrome X, 61, 63, 96,
101, 102, 103, 104, 105,
106, 107, 108, 111, 113,
225, 232, 255, 256, 293

**systemic lupus
erythromatosus** 160, 262

T3 18, 21, 258

T4 21

temperature 21, 197

thiols 214, 270

torture chamber 11, 13, 14,
19, 51, 79, 81, 82, 83, 95,
96, 98, 99, 133, 140, 141,
153

toxins 136, 137, 199, 209,
212, 213, 214, 215, 216,
239, 253, 287

triglycerides 90, 98, 102,
163, 167, 168, 169, 238,
240, 256, 257, 265

tyrosine 66, 69, 120, 275,
278, 279

ulcerative colitis 262

urine 35, 37, 39, 83, 85,
180, 181, 182, 189, 216,
223, 234, 276

vitamin A 19, 169, 211,
214, 240

vitamin B1, 214

vitamin B2 214

vitamin B3 214

vitamin B6 214, 279

zinc 124, 136, 192, 201, 214

About the Author

William B. Ferril M.D., has a bachelor of science in biochemistry and received his doctorate in medicine at UC-Davis, California. He completed his post-graduate education at Sacred Heart Medical Center in Spokane, Washington.

For most of the past eighteen years, Dr. Ferril practiced in Montana on the Flathead Indian Reservation, where he gained experience on the topics in this book. He works with chiropractors, naturopathic doctors, acupuncturists, and homeopaths. Two of his special interests are herbal and organic farming. In 2003, Dr. Ferril published *The Body Heals*.

Dr. Ferril's wife, Brenda, received her doctorate at Western States Chiropractic College in Portland, Oregon. They live with their son Conner in western Montana.

Let your patients discover how holistic medicine provides affordable, effective, and healing treatment.

To order copies of this book or of *The Body Heals,* fill out and send the following order form.

Name_____

Address_____

City_____State_____Zip_____

Phone_____

Quantity

_____Glandular Failure-Caused Obesity x $42.

_____The Body Heals x $50.

Shipping ($7. per book) $ _____

Purchase both books for $75.

Total $_____

We accept checks, credit cards, or money orders.

Send to:
The Bridge Medical Publishers
P.O. Box 324
Whitefish, MT 59937

Or Call: 406 863-9906
www.thebodyheals.com info@thebodyheals.com

Glandular Failure-Caused Obesity